IFIP Advances in Information and Communication Technology

612

IFIP – The International Federation for Information Processing

IFIP was founded in 1960 under the auspices of UNESCO, following the first World Computer Congress held in Paris the previous year. A federation for societies working in information processing, IFIP's aim is two-fold: to support information processing in the countries of its members and to encourage technology transfer to developing nations. As its mission statement clearly states:

IFIP is the global non-profit federation of societies of ICT professionals that aims at achieving a worldwide professional and socially responsible development and application of information and communication technologies.

IFIP is a non-profit-making organization, run almost solely by 2500 volunteers. It operates through a number of technical committees and working groups, which organize events and publications. IFIP's events range from large international open conferences to working conferences and local seminars.

The flagship event is the IFIP World Computer Congress, at which both invited and contributed papers are presented. Contributed papers are rigorously refereed and the rejection rate is high.

As with the Congress, participation in the open conferences is open to all and papers may be invited or submitted. Again, submitted papers are stringently refereed.

The working conferences are structured differently. They are usually run by a working group and attendance is generally smaller and occasionally by invitation only. Their purpose is to create an atmosphere conducive to innovation and development. Refereeing is also rigorous and papers are subjected to extensive group discussion.

Publications arising from IFIP events vary. The papers presented at the IFIP World Computer Congress and at open conferences are published as conference proceedings, while the results of the working conferences are often published as collections of selected and edited papers.

IFIP distinguishes three types of institutional membership: Country Representative Members, Members at Large, and Associate Members. The type of organization that can apply for membership is a wide variety and includes national or international societies of individual computer scientists/ICT professionals, associations or federations of such societies, government institutions/government related organizations, national or international research institutes or consortia, universities, academies of sciences, companies, national or international associations or federations of companies.

More information about this series at http://www.springer.com/series/6102

Gilbert Peterson · Sujeet Shenoi (Eds.)

Advances in Digital Forensics XVII

17th IFIP WG 11.9 International Conference
Virtual Event, February 1–2, 2021
Revised Selected Papers

 Springer

Editors
Gilbert Peterson
Department of Electrical
and Computer Engineering
Air Force Institute of Technology
Wright-Patterson AFB, OH, USA

Sujeet Shenoi
Tandy School of Computer Science
University of Tulsa
Tulsa, OK, USA

ISSN 1868-4238   ISSN 1868-422X (electronic)
IFIP Advances in Information and Communication Technology
ISBN 978-3-030-88383-6   ISBN 978-3-030-88381-2   (eBook)
https://doi.org/10.1007/978-3-030-88381-2

This Springer imprint is published by the registered company Springer Nature Switzerland AG
The registered company address is: Gewerbestrasse 11, 6330 Cham, Switzerland

Contents

Contributing Authors

Gunnar Alendal is a Special Investigator with Kripos/NCIS Norway, Oslo, Norway; and a Ph.D. student in Computer Security at the Norwegian University of Science and Technology, Gjovik, Norway. His research interests include digital forensics, reverse engineering, security vulnerabilities, information security and cryptography.

Shengqin Ao is an M.S. student in Cyber Security at the Institute of Information Engineering, Chinese Academy of Sciences, Beijing, China. Her research interests include threat intelligence information processing and information extraction.

Stefan Axelsson is an Associate Professor of Digital Forensics at the Norwegian University of Science and Technology, Gjovik, Norway; and a Professor of Digital Forensics and Cyber Security at Stockholm University, Stockholm, Sweden. His research interests include digital forensics, data analysis and digital investigations.

Harald Baier is a Professor of Digital Forensics at Bundeswehr University, Munich, Germany. His research interests include bulk data handling in digital forensics, data synthesis and RAM forensics.

Daniel Bastos is a Senior Cyber Security Engineer at Bosch Security and Safety Systems, Ovar, Portugal. His research interests include Internet of Things security, cloud security and information security and privacy.

Saheb Chhabra is a Ph.D. student in Computer Science and Engineering at Indraprastha Institute of Information Technology, New Delhi, India. His research interests include image processing and computer vision, and their applications in document fraud detection.

Kam-Pui Chow, Chair, IFIP WG 11.9 on Digital Forensics, is an Associate Professor of Computer Science at the University of Hong Kong, Hong Kong, China. His research interests include information security, digital forensics, live system forensics and digital surveillance.

Quang Anh Dang is an M.S. student in Information Technology Security at the Technical University of Darmstadt, Darmstadt, Germany. His research interests include penetration testing, the darknet and automobile security.

Marietheres Dietz is a Ph.D. student in Business Information Systems at the University of Regensburg, Regensburg, Germany. Her research focuses on the digital twin paradigm with an emphasis on security.

Geir Olav Dyrkolbotn is a Major in the Norwegian Armed Forces, Lillehammer, Norway; and an Associate Professor of Cyber Defense at the Norwegian University of Science and Technology, Gjovik, Norway. His research interests include cyber defense, reverse engineering, malware analysis, side-channel attacks and machine learning.

Ludwig Englbrecht is a Ph.D. student in Business Information Systems at the University of Regensburg, Regensburg, Germany. His research interests include new approaches in digital forensics and digital forensic readiness.

Pulkit Garg is a Ph.D. student in Computer Science and Engineering at the Indian Institute of Technology Jodhpur, Karwar, India. His research interests include image processing and computer vision, and their applications in document fraud detection.

Thomas Göbel is a Ph.D. student in Computer Science and a Researcher in the Research Institute of Cyber Defense at Bundeswehr University, Munich, Germany. His research interests include digital forensics, data synthesis and machine learning.

Garima Gupta is a Postdoctoral Researcher in Computer Science and Engineering at Indraprastha Institute of Information Technology, New Delhi, India. Her research interests include image processing and computer vision, and their applications in document fraud detection.

Gaurav Gupta, Vice Chair, IFIP WG 11.9 on Digital Forensics, is a Scientist E in the Ministry of Electronics and Information Technology, New Delhi, India. His research interests include mobile device security, digital forensics, web application security, Internet of Things security and security in emerging technologies.

Monika Gupta received her Ph.D. degree in Physics from the National Institute of Technology, Kurukshetra, India. Her research interests include image processing and computer vision and their applications in document fraud detection.

Weiqing Huang is a Professor of Cyber Security at the Institute of Information Engineering, Chinese Academy of Sciences, Beijing, China. His research interests include signal processing theory and technology, electromagnetic acoustic-optic detection and protection, and information security.

Jianguo Jiang is a Professor of Cyber Security at the Institute of Information Engineering, Chinese Academy of Sciences, Beijing, China. His research interests include network security, threat intelligence and data security.

Zhengwei Jiang is a Senior Engineer at the Institute of Information Engineering, Chinese Academy of Sciences, Beijing, China. His research interests include threat intelligence, malicious code analysis and suspicious network traffic analysis.

James Jones is an Associate Professor of Digital Forensics and Director of the Criminal Investigations and Network Analysis Center at George Mason University, Fairfax, Virginia. His research interests include digital artifact persistence, extraction, analysis and manipulation.

Gang Li is an Associate Professor of Information Technology at Deakin University, Burwood, Australia. His research interests include data science and business intelligence.

Myeong Lim received his Ph.D. degree in Information Technology from George Mason University, Fairfax, Virginia. His research interests include digital forensics, data mining and artificial intelligence.

Chao Liu is a Professor of Cyber Security at the Institute of Information Engineering, Chinese Academy of Sciences, Beijing, China. His research interests include mobile Internet security and network security evaluation.

Fucheng Liu is a Ph.D. student in Cyber Security at the Institute of Information Engineering, Chinese Academy of Sciences, Beijing, China. His research interests include insider threat detection and malicious entity detection.

Ning Luo is an M.S. student in Cyber Security at the Institute of Information Engineering, Chinese Academy of Sciences, Beijing, China. Her research interests include cyber security event extraction and threat intelligence analysis.

Yali Luo is an M.S. student in Cyber Security at the Institute of Information Engineering, Chinese Academy of Sciences, Beijing, China. Her research interests include threat intelligence analysis and threat intelligence assessment.

Suryadipta Majumdar is an Assistant Professor of Information Systems Engineering at Concordia University, Montreal, Canada. His research interests include cloud security, Internet of Things security, Internet of Things forensics and security auditing.

Günther Pernul is a Professor and Chair of Information Systems at the University of Regensburg, Regensburg, Germany. His research interests include information systems security, individual privacy and data protection.

Shengzhi Qin is a Ph.D. student in Computer Science at the University of Hong Kong, Hong Kong, China. His research interests include public opinion analysis, knowledge graphs and information security.

Anoop Singhal is a Senior Computer Scientist and Program Manager in the Computer Security Division at the National Institute of Standards and Technology, Gaithersburg, Maryland. His research interests include network security, network forensics, cloud security and data mining.

Nan Song is an M.S. student in Cyber Security at the Institute of Information Engineering, Chinese Academy of Sciences, Beijing, China. His research interests include malicious document detection, threat intelligence and data security.

Martin Steinebach is the Head of Media Security and Information Technology Forensics at the Fraunhofer Institute for Secure Information Technology, Darmstadt, Germany; and an Honorary Professor of Computer Science at the Technical University of Darmstadt, Darmstadt, Germany. His research interests include digital watermarking, robust hashing, steganalysis and multimedia forensics.

Changxin Su is an M.S. student in Computer Science at the Institute of Information Engineering, Chinese Academy of Sciences, Beijing, China. His research interests include threat intelligence information processing and information extraction.

Frieder Uhlig is an M.S. student in Information Technology Security at the Technical University of Darmstadt, Darmstadt, Germany. His research interests include network forensics and applications of approximate matching.

Qiaokun Wen is an M.S. student in Computer Science at the University of Hong Kong, Hong Kong, China. Her research interests include deep learning and information security.

Yu Wen is a Senior Engineer at the Institute of Information Engineering, Chinese Academy of Sciences, Beijing, China. His research interests include data mining, big data security and privacy, and insider threat detection.

Yanna Wu is an M.S. student in Cyber Security at the Institute of Information Engineering, Chinese Academy of Sciences, Beijing, China. Her research interests include threat perception detection and intelligent attack tracing.

Peian Yang is a Ph.D. student in Cyber Security at the Institute of High Energy Physics, Chinese Academy of Sciences, Beijing, China. His research interests include attack recognition and threat intelligence analysis.

York Yannikos is a Research Associate at the Fraunhofer Institute for Secure Information Technology, Darmstadt, Germany. His research interests include digital forensic tool testing, darknet marketplaces and open-source intelligence.

Min Yu is an Assistant Professor of Cyber Security at the Institute of Information Engineering, Chinese Academy of Sciences, Beijing, China. His research interests include malicious document detection, document content security and document security design and evaluation.

Preface

Digital forensics deals with the acquisition, preservation, examination, analysis and presentation of electronic evidence. Computer networks, cloud computing, smartphones, embedded devices and the Internet of Things have expanded the role of digital forensics beyond traditional computer crime investigations. Practically every crime now involves some aspect of digital evidence; digital forensics provides the techniques and tools to articulate this evidence in legal proceedings. Digital forensics also has myriad intelligence applications; furthermore, it has a vital role in cyber security – investigations of security breaches yield valuable information that can be used to design more secure and resilient systems.

This book, *Advances in Digital Forensics XVII*, is the seventeenth volume in the annual series produced by the IFIP Working Group 11.9 on Digital Forensics, an international community of scientists, engineers and practitioners dedicated to advancing the state of the art of research and practice in digital forensics. The book presents original research results and innovative applications in digital forensics. Also, it highlights some of the major technical and legal issues related to digital evidence and electronic crime investigations.

This volume contains thirteen revised and edited chapters based on papers presented at the Seventeenth IFIP WG 11.9 International Conference on Digital Forensics, a fully-remote event held on February 1-2, 2021. The papers were refereed by members of IFIP Working Group 11.9 and other internationally-recognized experts in digital forensics. The post-conference manuscripts submitted by the authors were rewritten to accommodate the suggestions provided by the conference attendees. They were subsequently revised by the editors to produce the final chapters published in this volume.

The chapters are organized into five sections: Themes and Issues, Approximate Matching Techniques, Advanced Forensic Techniques, Novel Applications and Image Forensics. The coverage of topics highlights the richness and vitality of the discipline, and offers promising avenues for future research in digital forensics.

This book is the result of the combined efforts of several individuals. In particular, we thank Kam-Pui Chow and Gaurav Gupta for their tireless work on behalf of IFIP Working Group 11.9 on Digital Forensics. We also acknowledge the support provided by the U.S. National Science Foundation, U.S. National Security Agency and U.S. Secret Service.

GILBERT PETERSON AND SUJEET SHENOI

I

THEMES AND ISSUES

Chapter 1

DIGITAL FORENSIC ACQUISITION KILL CHAIN – ANALYSIS AND DEMONSTRATION

Gunnar Alendal, Geir Olav Dyrkolbotn and Stefan Axelsson

Abstract   The increasing complexity and security of consumer products pose major challenges to digital forensics. Gaining access to encrypted user data without user credentials is a very difficult task. Such situations may require law enforcement to leverage offensive techniques – such as vulnerability exploitation – to bypass security measures in order to retrieve data in digital forensic investigations.

This chapter proposes a digital forensic acquisition kill chain to assist law enforcement in acquiring forensic data using offensive techniques. The concept is discussed and examples are provided to illustrate the various kill chain phases. The anticipated results of applying the kill chain include improvements in performance and success rates in short-term, case-motivated, digital forensic acquisition scenarios as well as in long-term, case-independent planning and research scenarios focused on identifying vulnerabilities and leveraging them in digital forensic acquisition methods and tools.

Keywords: Digital forensic acquisition, security vulnerabilities, kill chain

1.   Introduction

Several digital forensic process models have been proposed in the literature [2]. Regardless, a generic digital forensic process can be viewed as comprising four phases: seizure, acquisition, analysis and reporting. The digital forensic acquisition phase covers the retrieval of digital forensic data from seized devices and other data sources. Its main goal is to gain access to data for forensic analysis. Clearly, digital forensic acquisition tasks are changing as technology advances, but the overall goal is the same – accessing data in a forensically-sound manner [4, 7].

© IFIP International Federation for Information Processing 2021
Published by Springer Nature Switzerland AG 2021
G. Peterson and S. Shenoi (Eds.): Advances in Digital Forensics XVII, IFIP AICT 612, pp. 3–19, 2021.
https://doi.org/10.1007/978-3-030-88381-2_1

Embedded devices and online services are important sources of digital evidence in criminal cases, which makes digital forensic acquisition a priority for law enforcement. In recent years, smartphone vendors such as Apple and Samsung have instituted mechanisms for securing user data. Data in their devices is often encrypted and secured against a variety of attacks, local as well as remote. Gaining access to encrypted user data without user credentials is a very difficult task.

Garfinkel et al. [9] mention encryption as posing major challenges to law enforcement as they conduct digital forensic investigations. Arshad et al. [3] discuss the impacts of mandatory encryption and increased focus on privacy on the effectiveness of digital forensics. Balogun et al. [5] estimate that encryption alone prevents the recovery of digital forensic data in as much as sixty percent of cases that involve full disk encryption. In the FBI-Apple encryption dispute of 2015-16, Apple denied the FBI's request to create special firmware that would enable the recovery of user credentials from an iPhone 5C seized in a terrorist investigation [3]. Apple considered product security and user privacy to be more important than supporting the terrorism investigation.

Since law enforcement cannot rely on assistance from vendors to bypass security mechanisms in their products, the best option is to leverage offensive techniques to retrieve protected data in digital forensic investigations. Specifically, it is necessary to apply sophisticated techniques to discover published (n-day) and unpublished (0-day) vulnerabilities in the targets, and exploit them to acquire forensic data.

The idea of law enforcement leveraging published vulnerabilities is a concern because law enforcement assumes the role of an attacker in order to pursue justice. However, discovering and holding on to undocumented vulnerabilities in order to bypass security mechanisms are even more concerning. New vulnerabilities should be reported promptly to the affected vendors to enable them to mitigate risks, but this would prevent the continued use of the vulnerabilities. The conflicting interests between offensive and defensive uses of security vulnerabilities are not new. Indeed, they have been discussed publicly [8] and are addressed by the U.S. Government [21]. Whether to restrict discovered vulnerabilities for offensive use or disclose them for defensive purposes is determined by a vulnerability equities process, where U.S. agency representatives gather to evaluate and decide the fate of new vulnerabilities discovered by government agencies [21]. This policy is understandably controversial [19, 20].

This research does not take a stand on the vulnerability equities dilemma. Rather, it seeks to inform law enforcement about the possibility of discovering vulnerabilities in electronic devices and leveraging them to acquire forensically-sound data in criminal investigations. It focuses on a methodical approach called the "digital forensic acquisition kill chain," which is based on the "intrusion kill chain" concept used in computer network defense [10]. The intrusion kill chain is a systematic process for targeting and engaging an adversary to achieve the desired security effects [10]. The digital forensic acquisition kill chain turns this around – it is a systematic process for law enforcement (acting as an adversary) to target electronic devices using offensive techniques to facilitate digital forensic acquisition.

Law enforcement has some advantages when developing and employing offensive techniques. These include access to resources as well as police authority (ability to seize devices). Unlike attackers, law enforcement may have the time to execute offensive actions and impose patch prevention. A seized device may be fully patched with no known vulnerabilities at the time of seizure. However, the same device becomes vulnerable in the future as n-day vulnerabilities are published and 0-day vulnerabilities are discovered. Since law enforcement can prevent seized devices from receiving updates, it can leverage both types of vulnerabilities in digital forensic acquisition.

2. Related Work

Several digital forensic process models that focus on practitioners and the use of digital evidence in court have been proposed. The Advanced Data Acquisition Model [1] addresses the needs of practitioners and the expectations of courts for formal descriptions of the processes undertaken to acquire digital evidence. Montasari [17] has proposed a standardized model that enables digital forensic practitioners to follow a generic approach that can be applied to incident response as well as criminal and corporate investigations. In an attempt to further address the need for a generic digital forensic investigation process for use in the three domains, Montasari et al. [18] have proposed the Standardized Digital Forensic Investigation Process Model that draws on existing models and renders them generic enough for wide applicability. However, although digital forensic investigative processes are discussed, neither the scope nor the details of key processes such as examination and analysis are provided.

The three models address the need for trustworthy and court-accepted methods and processes. The focus is on ensuring the reliability of digital evidence presented in court using formal, standardized processes. In contrast, the digital forensic model presented in this chapter differs substantially from the three models in that it concentrates on using offensive techniques for digital forensic acquisition. However, the proposed model will have to be augmented in the future to guide the development of trustworthy, court-accepted methods.

3. Digital Forensic Acquisition Kill Chain

The primary goal of the proposed digital forensic acquisition kill chain is to articulate a structured process for developing new digital forensic acquisition methods based on offensive techniques. It is intended to improve performance and success rates during the time-constrained, case-motivated development of digital forensic acquisition methods as well as during the long-term case-independent development of digital forensic acquisition methods that take into account trends in consumer adoption of technology.

3.1 Background

Hutchins et al. [10] have specified a kill chain model that describes the network intrusion phases employed by advanced adversaries, often referred to as advanced persistent threats. Engaging a model that describes adversarial intrusion phases to inform defensive postures reduces the likelihood of success on the part of attackers. Specifically, detecting patterns that are signs of a campaign supports proactive computer network defense. This is referred to as intelligence-driven computer network defense, where identifying intrusion patterns facilitates responses before compromise occurs. The kill chain phases specify the goals and content as an adversary goes from intelligence gathering on a potential target to achieving full compromise and the ultimate goal of penetrating the target (e.g., exfiltrating sensitive data). Such a model is required because advanced adversaries invest considerable intellectual and technical resources to penetrate high value targets. The kill chain paradigm has proven to be very valuable, and several new ideas and models have been proposed [6, 11–13, 16].

The intrusion kill chain of Hutchins et al. [10] is motivated by the U.S. military targeting doctrine that encompasses six phases: find, fix, track, target, engage and assess. They adapted the targeting doctrine to computer network intrusions by introducing new phases. The resulting kill chain phases are: reconnaissance, weaponization, delivery, exploitation, installation, command and control, and actions on objectives. This methodical way of describing the expected adversarial phases views computer network defense from the adversaries' perspectives, facilitating detection by predicting the subsequent phases and the ability to execute proactive defensive operations. The research described in this

Figure 1. Generic digital forensic acquisition needs.

chapter adapts the intrusion kill chain to facilitate offensive actions in digital forensic acquisition scenarios.

3.2 Kill Chain Overview

Figure 1 shows a simplified view of digital forensic acquisition using offensive techniques. The proposed digital forensic acquisition kill chain adapts the original kill chain to specify a methodology for using offensive techniques in digital forensic acquisition, where law enforcement assumes the role of the adversary and seized devices (evidence containers) are the targets. It brings an intelligence-driven perspective to applying forensic data acquisition methods as well as researching and developing new methods.

Figure 2 shows the nine phases of the proposed digital forensic acquisition kill chain. The phases are: reconnaissance, identification, surveillance and vulnerability research, weaponization, delivery, exploitation, installation, command and control, and actions on objectives.

The nine phases are grouped and generalized according to the digital forensic acquisition needs in Figure 1. The initial reconnaissance phase considers the target of digital forensic acquisition. The next two phases, identification, and surveillance and vulnerability research, focus on the discovery of possible digital forensic acquisition solutions (vulnerabilities). The weaponization and delivery phases cover the development and realization of the discovered vulnerabilities. The last four phases,

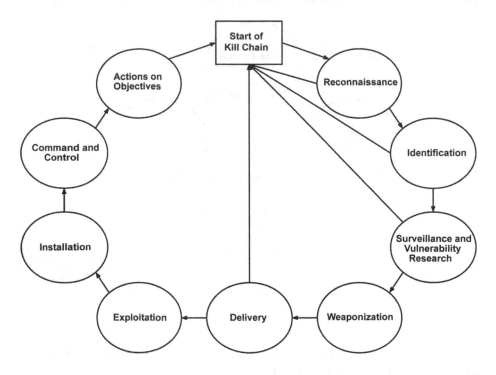

Figure 2. Digital forensic acquisition kill chain phases.

exploitation, installation, command and control, and actions on objectives, deal with operational issues.

A digital forensic acquisition kill chain is spawned in two general scenarios:

- **Case-Motivated Scenario:** This scenario is driven by a case-motivated need for a digital forensic acquisition method targeting a specific entity (e.g., device or service). Because digital forensic investigations are event-driven, law enforcement may not have applicable methods or be able to predict applicable methods for all possible scenarios. The kill chain focuses on solving the concrete challenge of acquiring forensically-sound data from the device or service, but it may spawn new kill chains to solve the sub-challenges that materialize. Several kill chains could be spawned in parallel and resources moved back and forth between them as the case foci and priorities change. The overall goal of a case-motivated kill chain is to apply digital forensic acquisition to a specific device or service.

- **Case-Independent Scenario:** This scenario is driven by a case-independent, intelligence-driven need for a digital forensic acquisition method that addresses a class of challenges. As the results of several case-motivated kill chains are obtained, a trend in the challenges encountered, such as the encryption of user data, could spawn its own kill chain. A challenge in another kill chain phase, say exploitation, could spawn a separate kill chain that focuses entirely on the challenges encountered during exploitation. A challenge related to a class of devices (e.g., from a specific vendor) could spawn a vendor-specific kill chain. The vendor could be Apple or Samsung, and the targets could be smartphones, services or components such as processors and flash memory chips that are common to vendor products or services. The overall goal of a case-independent kill chain is to improve the performance of subsequent case-motivated kill chains by leveraging intelligence, knowledge, methods and tools.

Upon considering the general digital forensic acquisition needs in Figure 1, the completion of the reconnaissance, identification, surveillance and vulnerability research, or delivery phases could result in the kill chain being terminated. For example, as shown in Figure 2, a kill chain covering a trending device would terminate at the end of the delivery phase because no operational needs exist. Of course, the completion of a phase could initiate the next phase, or the phase could spawn a new kill chain.

The initial phases of reconnaissance and identification could be performed at the start of an investigation to set the direction of the investigation and prioritize resources. An initial kill chain could spawn several new (sub) kill chains that address specific devices and services. This would, of course, depend on the amount of resources available. Prioritization and resource management of the sub kill chains would be a continuous process as the investigation proceeds.

Kill chains can also be applied to trending challenges that are detached from concrete investigations. This is motivated by the fact that many current digital forensic acquisition challenges are too complex to be solved given the limited time and resources available for investigations. The kill chains would focus on longer term challenges that need dedicated resources and prioritization. The available resources would be put to best use at all times, even in the case of parallel kill chains where resources would be shifted between kill chains as priorities change and commonalities are discovered. The expected results are increased knowledge of trending challenges, increased security expertise and new digital forensic acquisition methods.

3.3   Kill Chain Phases

This section discusses the nine phases of the digital forensic kill chain in detail.

Reconnaissance.   The reconnaissance phase focuses on the collection of information that would support the selection and prioritization of devices and services. This phase should be kept short if it is used as part of a specific case, where it would concentrate on selection and prioritization, and the estimation of the likelihood of success of a digital forensic acquisition method. In a case-independent scenario, the reconnaissance phase is more openly defined and may choose to focus on any target device or service of interest.

Multiple kill chains are expected to be initiated and terminated during the reconnaissance phase. Also, a single kill chain may spawn several kill chains for the identified devices and services. The basic idea is that the reconnaissance phase is based on the available information and information that is obtained easily.

Identification.   The challenge to developing a new forensic acquisition method is approached in a bottom-up manner. The focus is on identifying forensic data of value and the layers of security features that may prevent its access (e.g., encryption could be the first layer to bypass).

Volatility of forensic data is always an issue. Embedded devices often keep log files and unencrypted app data in random access memory (RAM) only. Thus, the digital forensic acquisition method must take into account the fact that a device cannot be power cycled. Addressing this challenge follows a different path in the remaining phases and would require a separate kill chain.

Note that two challenges – encryption and volatility – have been identified during this phase. Thus, two kill chains would be created and resource allocation decisions have to be made to best address the challenges.

Surveillance and Vulnerability Research.   During the surveillance and vulnerability research phase, existing vulnerabilities, techniques, tools and services are investigated. Also, resources are allocated to discover new vulnerabilities. Conducting activities in parallel can be efficient with regard to time. However, in order to optimize resources and not reinvent existing vulnerabilities and methods, the following two sub-phases are recommended:

- **Sub-Phase 1:** This short intelligence sub-phase focuses on gathering information about the identified challenges from open and closed sources. The goal is to discover published vulnerabilities that are potential candidates for direct use or are avenues for new vulnerability research. The sub-phase should not focus on the resource-intensive task of rediscovering low-level details about potential vulnerabilities, but only collect and prioritize potential vulnerabilities based on the available information.

- **Sub-Phase 2:** This sub-phase focuses on the active search for tools, services and vulnerabilities to address the identified challenges. It would also include a separate vulnerability research effort to discover new vulnerabilities.

The surveillance and vulnerability research phase is divided into two sub-phases in order to have a lightweight first sub-phase with a short time frame and low human resource needs. The results provide a basis for allocating resources to the much more intensive second sub-phase.

The second sub-phase has the most uncertainty with regard to resource needs and likelihood of attaining the end goal of a digital forensic acquisition method. However, in the event of success, a method that leverages a new vulnerability would have a longer life span than a method based on a published vulnerability. As multiple kill chains would be executed simultaneously during this sub-phase, efficient management of resources is required.

An example of a new kill chain is the discovery of new vulnerabilities and the acquisition of knowledge about existing vulnerabilities. Information about fixed vulnerabilities may be found at vendor web sites and in change logs and published patches. Although the information about a patched vulnerability often lacks the detail needed to isolate and trigger the vulnerability, an experienced vulnerability researcher would be able to obtain the information in a reasonable period of time. This could be hours, days or months depending on the complexity of the technology and vulnerability. Additionally, since a vulnerability may not always be convertible to a successful exploit, it is necessary to research several vulnerabilities. Identifying and studying vulnerabilities, and developing exploits are time consuming; also, predicting the resource needs is difficult. Therefore, it is important to balance time, resources and success potential between discovering new vulnerabilities and rediscovering known vulnerabilities by studying patches.

Weaponization. Weaponization involves the development of a working exploit from a new or existing vulnerability, a task that can be complex and potentially unrealizable. The weaponization of a vulnerability is hard to generalize, but it can be similar to the software development cycle. The steps proceed from developing a proof of concept to creating a production-quality exploit chain with quality assurance that minimizes the chance of failure when applied to digital forensic acquisition. A crucial step is to ensure that the method is forensically sound and complies with the law and established digital forensic standards [15].

Efforts in the weaponization phase also need to consider the users of the digital forensic acquisition method, especially their levels of expertise and access to special equipment and tools. Other considerations include ease of use, access to updates and support. Additionally, it is important to be aware that the type and sensitivity of the vulnerability may limit the number of users and cases where it can be applied.

Delivery. The delivery phase focuses on developing the channel or channels for executing the weaponized exploit. These could be physical interfaces such as USB, SPI, JTAG, UART and I2C or wireless channels such as Wi-Fi, Bluetooth and near-field communications (NFC). Even side channels that can be used to inject inputs into key components are potential delivery options.

Exploitation. During the exploitation phase, the focus is on applying the developed digital forensic acquisition method in a criminal case. Actions performed in this phase must adapt to the context of the device or service. Since the phase is operational in nature, it should consider all aspects of using the method, including device or service state, legality, special requirements, assumptions that do not hold (e.g., user credentials might be known), operational security and digital forensic principles. Special care should be taken if the exploitation is destructive (e.g., chip-off data acquisition), which would leave the device in a state where it cannot be returned to its owner after the investigation.

Installation. The installation phase is mostly concerned about the footprint required to achieve the goal of forensic data acquisition. A RAM-only installation is a good option when the goal is to acquire data from long-term storage and conform to forensically-sound principles [15]. Since a component installed in a device or service environment for forensic acquisition purposes may become a part of the acquired evidence, isolating and documenting the component and its behavior are vital in court proceedings. An alternative to installing a component is to enable device features to accomplish the same goal. For example, enabling `adb` and gaining root privileges on an Android smartphone would provide the required access. When executing custom code on a device, it is important that the footprint be as small as possible to reduce negative forensic impacts on volatile RAM storage. Alternatively, the available device debugging features could be leveraged.

Command and Control. In the command and control phase, control has already been gained over the execution and/or data on the target device or service. This could involve a generic interface such as a login shell with root access, arbitrary code execution or security feature (e.g., screen lock) bypass. Ideally, this phase should be detached from the earlier phases because it marks the start of the actual acquisition of digital forensic data. Activities could involve the use of special tools and commands that may not have been employed in the earlier device- or service-specific exploitation and installation phases. The advantage of separating command and control from other phases is the reuse of knowledge, code and tools. A login shell with root access may apply the same tools to acquire data from diverse Android devices, but activities in the exploitation and installation phases for Android devices from different vendors could be totally different and leverage completely different vulnerabilities to reach the command and control phase.

Actions on Objectives. The last phase in the kill chain is to simply execute the final goal of performing the digital acquisition to obtain data of forensic value from the device or service.

4. Case-Motivated Kill Chain Example

This section demonstrates the use of a digital forensic acquisition kill chain in a case-motivated scenario where law enforcement is interested in extracting data of forensic value from a broadband router seized at a crime scene. The data could constitute log files with network activity, including Wi-Fi logs pertaining to connected devices during a specific time period. The data could be used to gain information about the connected devices that would be identified by their MAC addresses. The device is a Zyxel router (model no. p8702n), which has a MIPS architecture and runs a uClinux-based operating system [14].

Reconnaissance. Open-source intelligence and reconnaissance activities for the Zyxel p8702n router focused on various discussion forums and on the availability of its firmware, which was eventually downloaded from a server located at `stup.telenor.net/firmwares/cpe-zyxel-p8702n`. Two firmware files, `100AAJX13D0.bin` and `100AAJX14D0.bin`, were obtained along with their `README` files.

Because change logs often contain valuable information about security patches, older files that were present on the server were also sought. The older files were downloaded from `web.archive.org`.

Thus, the reconnaissance phase yielded useful information from public forums along with publicly-available firmware files and their change logs.

Identification.   Forensic data with the most value was expected to reside in the flash memory of the Zyxel p8702n router. However, like many low-end embedded devices, the Zyxel p8702n router stores much of its data, including logs, in RAM only. This means that valuable forensic data could be lost if the device were to be turned off. This discovery is important because it impacts how the device should be seized; specifically, the device should not be powered down before digital forensic acquisition. Addressing the RAM memory acquisition challenge requires a separate kill chain.

Thus, two directions have to be pursued and a decision must be made about where to focus the available resources. The RAM data was assumed to be more valuable, so the corresponding kill chain was pursued – gaining access to the Zyxel p8702n router RAM data without turning off or restarting the device.

Surveillance and Vulnerability Research.   The shorter intelligence phase (sub-phase 1) sought to obtain information about acquiring RAM data, possibly by exploiting a vulnerability. In the case of the Zyxel p8702n router, a valuable source for vulnerability information was determined to be the vendor's patch reports. Because older firmware files and change logs were available, a reasonable approach was to examine the change logs for hints of security issues.

The examination revealed that firmware version `100AAJX7D0` had major security fixes. Therefore, the previous firmware version `100AAJX5D0` was inferred to have the security vulnerabilities.

The focus of the complex and resource-demanding sub-phase 2 was to rediscover the vulnerabilities patched in firmware version `100AAJX7D0`. This required firmware versions `100AAJX5D0` and `100AAJX7D0` to be unpacked and the differences between the two versions to be identified.

The analysis revealed a difference in the boot sequence, where a critical security vulnerability was exposed in the older version by a login shell on a serial console. The problem was that the login process `/bin/smd` had an SIGTSTP vulnerability – when `Ctrl-Z` was entered on the console, a `/bin/sh` shell was provided with the same credentials as the `/init` process. This enabled root access to the Zyxel p8702n router.

Thus, sub-phase 2 of the surveillance and vulnerability research phase resulted in the rediscovery of a vulnerability. However, the vulnerability still had to be triggered.

Weaponization. During the weaponization phase, it was determined that the vulnerability was not particularly difficult to exploit. The vulnerability was exploited by accessing the Zyxel p8702n router console and sending the SIGTSTP signal by entering Ctrl-Z. Thus, the goal of the weaponization phase was to discover an access method to the serial console of the Zyxel p8702n router; in this case, via the UART interface on the circuit board. The key result is that this could be done without powering off the Zyxel p8702n router.

Delivery. The delivery phase was also relatively simple. It involved sending Ctrl-Z over the attached serial console to the Zyxel p8702n router. The delivery was performed via the UART protocol using a standard RS232-to-USB serial converter and a putty terminal emulator.

Exploitation. Since the Zyxel p8702n router had to be powered on at all times, the digital forensic acquisition had to be performed without power-cycling the device. The considerations during the exploitation phase involved the ease of physical access to the device, speed of the operation (especially if it had to be covert), risk and likelihood of failure.

Important operational decisions had to made during the exploitation phase to prevent *ad hoc* decision making during the subsequent phases. Since the objective was to acquire data from RAM, any actions performed on the device (even as root) would affect the RAM (e.g., potentially overwriting valuable freed memory in RAM). Therefore, a bare minimum footprint had to be maintained.

Installation. The installation was restricted to digital forensic acquisition. Persistent access did not have to be maintained after the serial interface was detached. Therefore, no other tools were installed.

Command and Control. Root access to the Zyxel p8702n router rendered the digital forensic acquisition goal within reach. The command and control phase determined that only a few commands would be executed using on-device tools to preserve RAM content.

Actions on Objectives. At this point, all the digital forensic acquisition challenges were isolated and addressed. The final phase merely involved the actual digital forensic acquisition of RAM data in the Zyxel p8702n router.

Note that the primary goal was to focus on the raw RAM in order to preserve freed memory data and structures. Since this goal was achieved, it was not necessary to pursue the lower priority goal focusing on temporary RAM-only filesystems that are common in many Linux distributions, or the even lower priority goal focusing on flash memory.

5. Conclusions

Criminal investigations are increasingly hindered by strong security mechanisms that prevent forensically-relevant data from being acquired from electronic devices and services. Absent technical assistance from vendors and service providers, the only option for law enforcement is to leverage offensive techniques such as vulnerability exploitation to bypass security measures and acquire evidentiary data. The notion of law enforcement becoming an attacker in order to pursue justice is controversial, but police authority and search and seizure laws and regulations may support such actions.

The digital forensic acquisition kill chain described in this chapter adapts the kill chain employed in computer network defense to articulate a systematic methodology for using offensive techniques in digital forensic acquisition, where law enforcement assumes the role of the adversary and the seized devices and services of interest (evidence containers) are the targets. Applying the digital forensic acquisition kill chain provides many benefits – improvements in performance and success rates in short-term, case-motivated, forensic data acquisition scenarios as well as in long-term case-independent, intelligence-driven planning and research scenarios focused on identifying vulnerabilities and leveraging them in the development of novel digital forensic acquisition methods and tools.

Future research will focus on validating the digital forensic acquisition kill chain. The case study described in this chapter focused on a single device. Realistic field evaluations with diverse and more complicated challenges will provide valuable guidance on adjusting the kill chain phases. At this time, a single kill chain model has been proposed for case-motivated and case-independent scenarios. These scenarios appear to pull the kill chain model in different directions. As a result, future research will focus on creating separate digital forensic acquisition kill chain models for the two types of scenarios.

Acknowledgement

This research was supported by the IKTPLUSS Program of the Norwegian Research Council under R&D Project Ars Forensica Grant Agreement 248094/O70.

References

[1] R. Adams, V. Hobbs and G. Mann, The Advanced Data Acquisition Model (ADAM): A process model for digital forensic practice, *Journal of Digital Forensics, Security and Law*, vol. 8(4), pp. 25–48, 2012.

[2] A. Al-Dhaqm, S. Razak, R. Ikuesan, V. Kebande and K. Siddique, A review of mobile forensic investigation process models, *IEEE Access*, vol. 8, pp. 173359–173375, 2020.

[3] H. Arshad, A. bin Jantan and O. Abiodun, Digital forensics: Review of issues in scientific validation of digital evidence, *Journal of Information Processing Systems*, vol. 14(2), pp. 346–376, 2018.

[4] R. Ayers, S. Brothers and W. Jansen, Guidelines on Mobile Device Forensics, NIST Special Publication 800-101, Revision 1, National Institute of Standards and Technology, Gaithersburg, Maryland, 2014.

[5] A. Balogun and S. Zhu, Privacy impacts of data encryption on the efficiency of digital forensics technology, *International Journal of Advanced Computer Science and Applications*, vol. 4(5), pp. 36–40, 2013.

[6] S. Caltagirone, A. Pendergast and C. Betz, The Diamond Model of Intrusion Analysis, Technical Report ADA586960, Center for Cyber Threat Intelligence and Threat Research, Hanover, Maryland, 2013.

[7] E. Casey, *Digital Evidence and Computer Crime: Forensic Science, Computers and the Internet*, Elsevier, Waltham, Massachusetts, 2011.

[8] M. Daniel, Heartbleed: Understanding when we disclose cyber vulnerabilities, *White House Blog*, The White House, Washington, D.C. (obamawhitehouse.archives.gov/blog/2014/04/28/heartbleed-understanding-when-we-disclose-cyber-vulnerabilities), April 28, 2014.

[9] S. Garfinkel, Digital forensics research: The next 10 years, *Digital Investigation*, vol. 7(S), pp. S64–S73, 2010.

[10] E. Hutchins, M. Cloppert and R. Amin, Intelligence-driven computer network defense informed by analysis of adversary campaigns and intrusion kill chains, in *Leading Issues in Information Warfare and Security Research*, J. Ryan (Ed.), Academic Publishing, Reading, United Kingdom, pp. 80–106, 2011.

[11] G. Ioannou, P. Louvieris, N. Clewley and G. Powell, A Markov multi-phase transferable belief model: An application for predicting data exfiltration APTs, *Proceedings of the Sixteenth International Conference on Information Fusion*, pp. 842–849, 2013.

[12] M. Khan, S. Siddiqui and K. Ferens, A cognitive and concurrent cyber kill chain model, in *Computer and Network Security Essentials*, K. Daimi (Ed.), Springer, Cham, Switzerland, pp. 585–602, 2018.

[13] R. Luh, M. Temper, S. Tjoa and S. Schrittwieser, APT RPG: Design of a gamified attacker/defender meta model, *Proceedings of the Fourth International Conference on Information Systems Security and Privacy*, pp. 526–537, 2018.

[14] D. McCullough, uCLinux for Linux programmers, *Linux Journal*, vol. 2004(123), article no. 7, 2004.

[15] R. McKemmish, When is digital evidence forensically sound? in *Advances in Digital Forensics IV*, I. Ray and S. Shenoi (Eds.), Springer, Boston, Massachusetts, pp. 3–15, 2008.

[16] B. Messaoud, K. Guennoun, M. Wahbi and M. Sadik, Advanced persistent threat: New analysis driven by life cycle phases and their challenges, *Proceedings of the International Conference on Advanced Communications Systems and Information Security*, 2016.

[17] R. Montasari, A standardized data acquisition process model for digital forensic investigations, *International Journal of Information and Computer Security*, vol. 9(3), pp. 229–249, 2017.

[18] R. Montasari, R. Hill, V. Carpenter and A. Hosseinian-Far, The Standardized Digital Forensic Investigation Process Model (SDFIPM), in *Blockchain and Clinical Trial*, H. Jahankhani, S. Kendzierskyj, A. Jamal, G. Epiphaniou and H. Al-Khateeb (Eds.), Springer, Cham, Switzerland, pp. 169–209, 2019.

[19] T. Moore, A. Friedman and A. Procaccia, Would a "cyber warrior" protect us? Exploring trade-offs between attack and defense of information systems, *Proceedings of the New Security Paradigms Workshop*, pp. 85–94, 2010.

[20] B. Schneier, Disclosing vs. hoarding vulnerabilities, *Schneier on Security Blog* (`www.schneier.com/blog/archives/2014/05/disclosing_vs_h.html`), May 22, 2014.

[21] The White House, Vulnerabilities Equities Policy and Process for the United States Government, Washington, D.C. (`trumpwhitehou se.archives.gov/sites/whitehouse.gov/files/images/Exter nal-UnclassifiedVEPCharterFINAL.PDF`), November 15, 2017.

Chapter 2

ENHANCING INDUSTRIAL CONTROL SYSTEM FORENSICS USING REPLICATION-BASED DIGITAL TWINS

Marietheres Dietz, Ludwig Englbrecht and Günther Pernul

Abstract   Industrial control systems are increasingly targeted by cyber attacks. However, it is difficult to conduct forensic investigations of industrial control systems because taking them offline is often infeasible or expensive. An attractive option is to conduct a forensic investigation of a digital twin of an industrial control system. This chapter demonstrates how a forensic investigation can be performed using a replication-based digital twin. A digital twin also makes it possible to select the appropriate tools for evidence acquisition and analysis before interacting with the real system. The approach is evaluated using a prototype implementation.

Keywords: Digital forensics, industrial control systems, digital twins

1.   Introduction

Industrial control systems have long life-spans. Since maintenance is performed only a few times a year, industrial control system firmware and software are updated very infrequently [19]. While safe operations are a priority for industrial control systems, adequate levels of security are generally lacking. As a result, industrial control systems are exposed to numerous threats.

When security is breached, an incident response is initiated to understand the situation, mitigate the effects, perform corrective actions and ensure safe operations. A digital forensic investigation provides the best insights into an incident and also assists in the prosecution of the perpetrators. Digital forensic readiness is essential to maximize the ability

© IFIP International Federation for Information Processing 2021
Published by Springer Nature Switzerland AG 2021
G. Peterson and S. Shenoi (Eds.): Advances in Digital Forensics XVII, IFIP AICT 612, pp. 21–38, 2021.
https://doi.org/10.1007/978-3-030-88381-2_2

to acquire useful evidence and minimize the costs of investigations [30]. An appropriate enterprise-wide maturity level is needed to implement digital forensic readiness for information technology assets [13]. However, an appropriate maturity level is even more difficult to attain by enterprises with operational technology assets such as industrial control systems.

Conducting digital forensic investigations of industrial control systems is challenging because the systems are required to operate continuously for safety and financial reasons. Since the systems cannot be stopped to acquire evidence and conduct forensic analyses, the only alternative is to reduce the shutdown time. This can be accomplished using digital twins of the real systems to identify where evidence resides in the real systems and to select the right tools for extracting evidence before the real systems are stopped. Digital twins replicate the dynamic behavior of their real counterparts. Unlike other state-of-the-art solutions, using digital twins enable industrial control systems to continue to operate while potential attacks are being investigated. Furthermore, unlike cyber ranges and testbeds, digital twins are well suited to digital forensics due to their fidelity, flexibility and two-way communications between the real systems and their digital twins. This chapter demonstrates how forensic investigations of industrial control systems can be performed using replication-based digital twins.

2.   Background

This section provides an overview of digital twins, digital twin security and digital forensics.

2.1   Digital Twin

A digital twin is a controversial term with different meanings in different domains [23]. Nevertheless, it can be regarded as a virtual representation of a real object over its lifecycle. Although digital twins have been employed in several domains, including smart cities [14], healthcare [21] and product management [31], they are commonly deployed in the Industry 4.0 paradigm [23], which is the focus of this work.

According to Kritzinger et al. [20], a digital twin is distinguished from other virtual representations (e.g., digital models) by its data flow. A digital model is a manual flow from a real object to a digital object. In contrast, a digital twin has bidirectional automated data flows between the real and virtual worlds [4, 20]. Thus, the digital twin is able to gather state data from its physical counterpart. However, the twin usually contains other asset-relevant data such as specification data [4]. When enhanced with semantics [26], this data can support various analyses, optimizations and simulations performed by the digital twin [1, 16].

2.2 Digital Twin Security

Several researchers have emphasized that digital twins must have adequate security [16, 26]. However, digital twins can also support industrial control system security [5, 8, 25]. Thus, digital twin security has two perspectives, securing digital twins and using digital twins to implement security. This work focuses on the second perspective and assumes that a digital twin has adequate security. The following security-centered definition of a digital twin is employed in this work [9].

Definition 1. A digital twin is a virtual double of a system during its lifecycle, providing sufficient fidelity for security measures via the consumption of real-time data when required.

Various digital twin modes exist to enable secure operations [5]. While a digital twin provides analytical and simulation capabilities, the replication mode, which supports the exact mirroring of a real system and its states, is relevant to this research [5]. The record and play functionality [10] is unique to the replication mode and is the essence of this work (Definition 1). While research has focused on digital twin security, little, if any, work has focused on using digital twins to support digital forensic investigations despite promising characteristics such as system state mirroring.

2.3 Digital Forensics

Digital forensics involves the identification, collection, preservation, validation, analysis, interpretation and presentation of digital evidence associated with an incident in a computer system [24]. The collection and analysis of digital evidence should be based on a comprehensive process model (see, e.g., [18]).

Internet of Things devices generate many traces during their operation that can be vital to digital forensic investigations. Clearly, there is a need for tools that can support digital forensic investigations of Internet of Things devices. Servida and Casey [27] discuss the challenges involved in examining Internet of Things devices. Although their work does not explicitly deal with industrial control systems, the three main challenges they present are relevant to this work. First, the computing power of the devices is very low and not suitable for performing complex tasks.

Second, most devices have limited embedded memory or an external SD card. Third, it is difficult to extract evidence due to device heterogeneity.

Digital forensic practitioners require considerable expertise, tools and time to completely and correctly reconstruct evidence given the large amounts of data to be processed. A promising approach is to use digital twins. A digital twin can be used to detect an attack. Additionally, it can provide crucial information and insights during digital forensic analysis.

Another challenge to conducting a digital forensic analysis is that the integrity of the data can be compromised during its recovery and analysis. A digital twin can assist in ensuring data integrity. The digital twin of an industrial control system can be examined and the digital forensic process and results verified before performing any actions on the real system.

3.   Related Work

The application of digital twins to security and especially forensics has only recently drawn the attention of researchers. The concepts proposed by Eckhart et al. [7, 10] and Gehrmann and Gunnarsson [15] are closely related to this research.

In general, a digital twin is a high-fidelity representation of its real counterpart. Eckhart and Ekelhart [7] were the first to study industrial control system state replication using digital twins; their focus was on reflecting the states of the real system virtually. To avoid large bandwidth overhead, Eckhart and Ekelhart [7] proposed a passive approach that identifies stimuli that alter real-world system states and reproduces them in the digital world.

Gehrmann and Gunnarsson [15] demonstrated that an active approach is suitable for less complex digital twins with moderate synchronization frequencies that do not create overhead; examples are replications of a single industrial control system or an industrial plant with a few industrial control systems. They showed that synchronizing the states of a real system with a digital twin supports active replication. Gehrmann and Gunnarsson also specified security requirements, established a security architecture and implemented secure synchronization between the real object and its digital twin.

According to Eckhart et al. [10], the record and play (replay) mode is a special manifestation of replication using a digital twin. Generally, a digital twin would reflect the real system states at all times. Additionally, restoring the preceding state is enabled, instead of merely replicating the current state that would be lost as soon as the subsequent state is replicated. The replay mode supports incident management, explicitly tracks infection histories [9] and provides novel functionality with regard to forensics [5]. Therefore, replay is a vital functionality provided by replication using digital twins, especially when applied to digital forensics.

Modern industrial control systems are exposed to security threats because they incorporate commodity hardware and software and operate in highly-interconnected environments. Researchers have attempted to enhance digital forensic capabilities by providing monitoring and logging functionality [3, 33].

The proposed research differs from the research described above in a key way. If an industrial control system is compromised, then its digital twin would exhibit the same malicious behavior as its real counterpart. While this is often considered a downside of replication [7, 15], the proposed research attempts to transform it to an advantage. Specifically, after replicating the exact states of the real system in its digital twin, forensic tools can be applied to conduct deep inspections of a security incident.

Note that this approach differs from traditional intrusion detection and security incident and event management (SIEM) research by focusing on deep inspection and resolution of incidents instead of mere detection. Indeed, the objective is to create a forensically-sound and replicable baseline for forensic analyses of industrial control systems.

The proposed approach advances the state-of-the-art in several respects. Artificial environments such as Mininet can be used to reproduce stimuli (or events) that change the states of industrial control systems or are responsible for the identified activities [7]. Such environments are well-suited to simulating systems and their events [8], but the fidelity of their replication is reduced [7]. Therefore, the proposed approach includes stimuli and events directly from a real system. It is also relevant to note that, while the digital-twin-based security framework of Gehrmann and Gunnarsson [15] considers security abstractly by suggesting a single security analysis component for multiple digital twins in a system, the proposed approach incorporates a separate security analysis module for each control system.

The proposed approach also advances previous work by providing a concrete definition and a proof-of-concept implementation of a digital twin environment for forensic (and/or security) analyses. Instead of investigating how digital twins can protect industrial control systems from external attacks [15], the focus is on digital forensics in the factory domain. The replication-based approach incorporates real communications between multiple control systems. The influences on the logical compo-

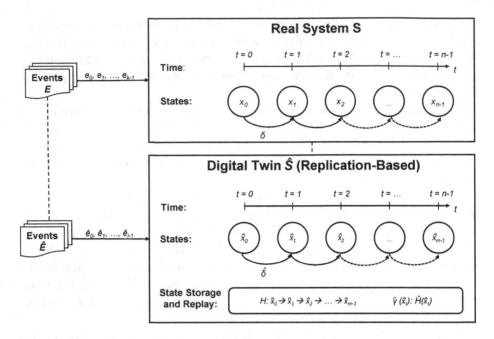

Figure 1.   Replication-based digital twin with state storage and replay.

nents and filesystems of industrial control systems are also mimicked. Whereas other work relies on logging and monitoring mechanisms [3, 33], the proposed approach focuses on filesystem changes by making periodic recordings of content in digital twins. This enables the replaying of all the evidence generated by the replicated digital twins.

4.   Replication Using Digital Twins

This section provides the theoretical foundations for the proposed approach using digital twins. Four theorems formalize the requirements for replication-based digital twins with replay capabilities.

4.1   Replication and Replay Theorems

Definition 1 above provides the security-centered definition of a digital twin [9]. Formal notions pertaining to state replication with digital twins are provided in [7, 15]. These notions and additional definitions create the basis for digital forensic analyses. Figure 1 presents the proposed replication-based digital twin with replay functionality that can be used for digital forensic analyses.

Sufficient fidelity of digital twins in the replication mode is required to support analyses. This concept is formalized by Theorems 1, 2 and 3.

Theorem 1 (Representation of States). *The finite set of states $X = \{x_0, x_1, ..., x_{n-1}\}$ of a real system is represented in its replication-based digital twin as $\hat{X} = \{\hat{x}_0, \hat{x}_1, ..., \hat{x}_{m-1}\}$. A high-fidelity digital twin is replicated as a subset of the real system corresponding to $\hat{X} \subseteq X$ where $m \leq n$. In an ideal digital twin, $\hat{X} = X$.*

Theorem 2 (Timely Orderliness). *To replicate a real system accurately, the concept of time has to be considered. Let $x_t \in X_t$ represent the real system at time t where the initial state is x_0. The digital twin replicates each state $\hat{x}_t$ in chronological order so that $x_0 < x_1 < x_2 < ... < x_{n-1}$. Time delays may occur between the real system states and digital twin states, but they do not affect digital forensics much because forensic investigations are typically conducted post mortem.*

In addition to having sufficient fidelity, a digital twin must be able to consume real-time data and replay it. This motivates Theorem 3.

Theorem 3 (Integration of Events.) *If a system changes from one state to another, certain input data is required, which is referred to as events. Events might occur due to the inner workings of the system or due to its external environment that may not be covered by its digital twin. For example, commands from the system's program are internal events whereas network traffic from other systems are external events. Real events $E = \{e_0, e_1, ..., e_{k-1}\}$ and the events replicated in the digital twin $\hat{E} = \{\hat{e}_0, \hat{e}_1, ..., \hat{e}_{l-1}\}$ express these inputs, where $\hat{E} \subseteq E$ and $l \leq k$. Furthermore, the transition function δ expresses the changes of states in the real system: $\delta : X \cdot E \to X$, i.e., $x_{t+1} = \delta(x_t, e_t)$. Likewise, $\hat{\delta}$ expresses the changes of states in its digital twin. Events lead to state changes that in turn may leave traces such as new files or updates of internal values in the real system. As a result of replication, the same traces will be found in the digital twin.*

The highly-desirable replication-based replay functionality imposes additional requirements. This motivates a fourth theorem.

Theorem 4 (Accuracy in Replay). *The replay function resets a digital twin to a starting state. Deviations from the previously-observed states of the digital twin should not occur when retrieving its historical events [9]. First, the transition function leading to a subsequent state x' is repli-*

cated to achieve similar states in the digital twin: $\delta(x,e) = \hat{\delta}(\hat{x},\hat{e}) \iff$ $x' = \hat{x}'$. *Thus, starting with the initial state, a chain of historic states* $\hat{H} : \hat{x}_0 \mapsto \hat{x}_1 \mapsto \hat{x}_2 \mapsto ... \mapsto \hat{x}_n$ *can be constructed. This chain can be used to reset states and replay the subsequent states in chronological order. The replay function* $\hat{\gamma}(\hat{x}_t) : \hat{H}(\hat{x}_t)$ *expresses a reset to state* $\hat{x}_t$ and the traversing of the states in chronological order.

Finally, the real system is a deterministic system defined by tuples $S := (X, x_0, E, \delta)$. The digital twin is represented similarly by incorporating state storage and replay functionality $\hat{S} := (\hat{X}, \hat{x}_0, \hat{E}, \hat{\delta}, \hat{H}, \hat{\gamma})$.

4.2   Conceptual Framework

The framework collects data (e.g., network traffic) from a real system that is imported by a high-fidelity digital twin of the real system. The replication ensures that the states of the real system are mirrored by the digital twin. Thus, the digital twin would have digital evidence traces that mirror those in the real system, enabling the digital twin to be analyzed in a forensic investigation.

Forensic investigations, however, require the recording of the current replicated states as well previous states and their associated traces. Furthermore, the states should be accountable and in chronological order (specified in Theorem 2). Therefore, the replication-based digital twin framework also incorporates state storage and replay functionality.

Figure 2 shows the replication-based digital twin framework with state storage and replay functionality. Note that the events $\hat{E}$ in the framework may be external as well as internal.

The framework has four key building blocks: (i) data collection, (ii) digital twin replication, (iii) digital twin state storage and replay, and (iv) digital forensic analysis:

- **Data Collection:** Input data is required to replicate a real system with sufficient fidelity. The data collection component gathers the inputs (events $\hat{E}$) as specified in Theorem 3. The input data may be internal (static) or external (dynamic). Internal data typically can be obtained directly from the real system, such as program code that may alter the system state or commands that are sent in response to external events. They are static because they do not change often and do not exhibit streaming characteristics. In contrast, external data typically corresponds to events that affect the real system. They often occur outside the real system, but within its environment. External data can be characterized as mainly dynamic because it can emerge at any time and in a constant manner

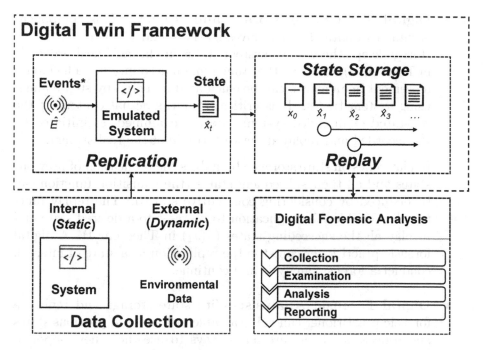

Figure 2. Digital twin framework.

(streaming characteristics). Examples of external data are network traffic from the real system's environment and sensor values.

- **Digital Twin Replication:** The collected data is used to replicate the real system as a digital twin. Internal data such as program code and system configurations are used to emulate the real system. It is important to choose the right technology for replication along with the system itself, desired degree of detail and levels of representation (e.g., network, software and operating system).

While internal events automatically occur by integrating program code and software, external events have to be input to the system. This enables the replication of system behavior based on internal events and on external stimuli. The greater the amount of data integrated, the more accurate the replication, but a trade-off should be performed between replication fidelity and cost. According to Theorems 1, 2 and 3, the states of the real system are replicated in the desired manner. In each state, different traces are created based on the transition functions. For instance, a file could be created in state A while its content is changed in state B.

- **Digital Twin State Storage and Replay:** This component
  is vital to digital forensic investigations and elucidation. Stage
  storage keeps the various system states in chronological order ($\hat{H}$
  in Theorem 4), ensuring that they remain accountable. The lateral
  movement of an attack can be elucidated in a step-by-step manner
  using system traces. It is critical that the actual content of the
  generated or modified system data are examined, as initiated by
  the recorded and replayed network traffic of the real system.

  Replay relies on state storage. It enables the replication of previous
  states and all their subsequent states (the transition function can
  be deduced by considering consecutive states). The replay functionality enables the replication to be reset to a desired state and
  to play all the succeeding states ($\hat{\gamma}(\hat{x}_t)$ in Theorem 4). A digital
  forensic practitioner can stop the replication at a state of interest,
  conduct a forensic analysis and continue.

- **Digital Forensic Analysis:** With state storage and replay, a
  forensic practitioner can gain valuable insights into previous states
  and processes. There are many ways to use these new opportunities in digital forensics. For example, a practitioner could reset
  the replication to a certain state using the replay functionality and
  use analysis tools in turn until a suitable tool is found. The exact
  replication of events and their storage in chronological order enhances the understanding of an attacker's tactics, techniques and
  procedures. With each consecutive state, the pattern of the attack
  might become clearer to the practitioner. An attack that deletes
  its traces can be investigated by leveraging the replication-based
  approach with state storage and replay according to the digital
  forensic process defined by Kent et al. [18].

5.   Implementation and Evaluation

The implementation involved a real system comprising a windmill
with a programmable logic controller (PLC). The real system and its
digital twin had identical Unix operating systems, ran the OpenPLC programmable logic controller software and communicated using TCP/IP.

5.1   Implementation and Experimental Setup

The real system was replicated by passively capturing its network traffic using `tshark` and re-transmitting the traffic to the emulated system.
The configuration supports forensic analyses of the real system during
execution. This section provides details of the implementation.

The implementation incorporated four main components: (i) data collection, (ii) digital twin replication, (iii) digital twin state storage and replay, and (iv) digital forensic analysis:

- **Data Collection:** The real system and its digital twin employed Unix operating systems running OpenPLC software. OpenPLC supports the IEC 61131-3 standard [17], which defines the software architecture and programming languages for programmable logic controllers. Network traffic created during the operation of the real system was recorded as a PCAP file and re-transmitted to the digital twin.

- **Digital Twin Replication:** The real system was replicated by the digital twin, which executed in a virtual environment running the same Unix operating system and OpenPLC software. All the OpenPLC variables in the real system were refactored as "memory storage" in the digital twin. This facilitated the persistent storage feature of OpenPLC whereby values at various programmable logic controller addresses were saved to disk to provide insights into their changes during programmable logic controller operation.

  The open-source software Polymorph [29] was used to re-transmit network traffic. Polymorph translated the PCAP file data to templates to enable situation-aware interactions. It also facilitated dynamic integration of the components instead of the pure transmission of data. The templates were modifiable for subsequent re-transmission, which was vital because the digital twin had to respond correctly to commands issued in the real system. The digital twin operated (according to the transition function in Theorem 3) as closely as possible to the real system.

- **Digital Twin State Storage and Replay:** The storage and replay component was hosted on a virtual machine (VM), which created a storage snapshot of system state every minute. The snapshots enabled the entire virtual system to be re-created at any point in time. Since data was written and deleted during the execution of OpenPLC, the freed data area in the digital twin could be overwritten, which was problematic. Therefore, a modified version of stateless continuous data protection (CDP) software was employed. Continuous data protection is a technology that continuously captures and stores data changes, enabling data from any point in the past to be recovered [28, 34]. The software also enables the monitoring and restoration of all files generated during system execution.

The `sauvegardeEx` tool [12] was created to gain insights into the generation of files during OpenPLC operation. It is based on `sauvegarde`, an open-source, stateless implementation of continuous data protection [6].

Compared with traditional backup technologies, continuous data protection mechanisms improve the recovery point objective (RPO) metric [22]. The RPO metric defines the time between two successful backups and, thus, the maximum amount of data loss during a successful recovery. The RPO is zero for a system with fully synchronized protection. The RPO metric of zero provided by continuous data protection theoretically allows unlimited recovery points [28]. The `sauvegardeEX` tool implements continuous data protection at the block level (virtual machine snapshots) and at the file level. The virtual machine snapshots were also taken every minute to recover the running system, including volatile RAM memory. With more resources, snapshots could be taken more frequently.

The digital twin framework, which was equipped with the client version of `sauvegardeEx`, sent every file alteration along with the file content to the server. This enabled a specific file to be restored at any point in time and also addressed the problem of overwriting a freed storage area (due to file deletion or update). Instituting this mechanism during OpenPLC execution and replicating the environment via Polymorph ensured that all possible traces were recorded.

■ **Forensic Analysis:** Continuous data protection was also exploited to generate data for digital forensic analysis. Continuous data protection technology has not been considered in the digital forensic context, but it is certainly important. In fact, the state storage and replay functionality supported by the modified continuous data protection mechanism enables different forensic tools to be used without the risk of compromising data in the real system or even rendering the data unusable. Indeed, the replication-based approach with state storage and replay completely supports the digital forensic process specified by Kent et al. [18].

5.2   Results and Evaluation

To evaluate the framework, network traffic between OpenPLC (master) and the control unit of the windmill (slave) was captured. The standard Modbus TCP protocol was used for communications. A Python-based Modbus simulation tool `pyModbusTCP` was used to generate realistic sensor data. A `sensordata.py` script simulated the sensor data for the wind speed around the windmill. Eight registers in the Modbus slave device were relevant. The first four registers contained the current sensor data and the other four indicated the corresponding system status. The sensor data values ranged from one to ten. The sensor values were grouped into three system state categories, green, yellow and red:

- **Values 1-5:** System state is green (Statuscode 200).

- **Values 6-8:** System state is yellow (Statuscode 300).

- **Values 9-10:** System state is red (Statuscode 400).

Pseudorandom sensor values were written to the registers every ten seconds by the Python script. Pseudorandom numbers were used so that changes to the wind speed corresponded to ascending or descending patterns and the values did not vary too much. OpenPLC updated the system status every five seconds based on the sensor values. A slight delay always occurred before the new system status was stored in the registers.

Traffic between OpenPLC and the Modbus slave was transformed to the network templates by Polymorph. This mimicked the external behavior based on traffic content.

By applying Polymorph and `sauvegardeEx` to the replicated system running OpenPLC with persistent storage, all the states of the OpenPLC addresses and related file changes during execution were recorded. All the system artifacts were simultaneously recorded at the digital twin. This enabled the determination of the relationships between the changed states and file content at any point during execution. Since the framework was designed to acquire the actual file content of written files on the hard drive as well as volatile memory content, VirtualBox and its live snapshotting functionality were leveraged.

During the evaluation, 1,432 state changes were observed on the persistent data storage during a five-minute period. The periodic virtual machine snapshots enabled a retrospective analysis of the running system.

Several forensic tools were applied at three points in time during execution. Data used for the evaluation is available at [11]. Table 1 shows an excerpt of the analysis and suitable digital forensic tools. The tools presented in [32] were considered in the evaluation.

An important component of the evaluation was to analyze discrepancies between the real system and its digital twin. This was accomplished by comparing file-level evidence between the real system and its digital

Table 1.   Excerpt of the recorded state changes and suitable digital forensic tools.

Timestamp	State Change	File Name	Suitable Tools
2020-09-11 09:16:59.123	File modification	persistent.file	CPLCD
2020-09-11 09:17:01.102	File modification	openplc.db	Bring2lite
2020-09-11 09:17:23.322	File modification	persistent.file	CPLCD
2020-09-11 09:18:00.000	VM snapshot	%/disk.vmdk	Autopsy, CPLCD
2020-09-11 09:18:01.202	File modification	persistent.file	CPLCD
2020-09-11 09:18:01.302	File modification	openplc.db	Bring2lite

twin using the approximate hashing function of Breitinger and Baier [2]. This hashing function was used instead of the SHA-256 hashing function because it provides measures of file similarity. Frequent comparisons of the recorded files helped identify and verify time-event correlations. Comparisons of the 1,432 recorded state changes yielded an average similarity of 98%.

6.   Discussion

The proposed approach is easily implemented on architectures with open-source programmable logic controller software. All that is needed is knowledge about the real system and adequate recordings of network traffic.

Although a digital twin adequately replicates a real system, this research has omitted formal measurements of the similarity between them. In order for evidence from a digital twin to be admissible, it is vital their similarity be measured and documented. One approach is to use the synchronization function proposed by Gehrmann and Gunnarsson [15]. However, this mechanism can introduce time differences between the digital twin and its real counterpart. Specifically, system states caused by an attacker in the real system would manifest themselves earlier than in the digital twin.

An interesting possibility is to incorporate control theory in a digital twin. This would make the digital twin a better replication of the real system that would, in turn, contribute to the admissibility of the extracted evidence.

The proposed approach provides recordings of file content at various points in time (via sauvegardeEx) and system-wide snapshotting of the running programmable logic controller software (via VirtualBox). These features make it possible to detect and analyze RAM-based malware. However, a limitation is that the implementation employed Unix and open-source OpenPLC software instead of industrial control system firmware. Although the underlying theory is sound, the open-source implementation would hinder its application in industrial environments.

The implementation of a digital twin for forensic investigations can be expensive. In addition to creating a digital twin and verifying its fidelity, it would be necessary to constantly modify the digital twin and verify that it keeps up with any and all changes made to the real system. This would require digital forensic professionals to have considerable industrial control system expertise, which would be an expensive proposition.

7. Conclusions

As attacks on critical infrastructure assets increase, it is imperative to develop digital forensic techniques targeted for industrial control systems. However, taking an industrial control system offline to conduct a digital forensic investigation is infeasible and expensive. An attractive alternative is to conduct a forensic investigation of a digital twin of an industrial control system. Implementing a digital twin with replication-based state storage and replay enables the acquisition and analysis of file-level evidence. Additionally, the digital twin could be used to select the appropriate forensic tools for evidence acquisition and analysis before interacting with the real system, thereby reducing system downtime when conducting the investigation.

Acknowledgement

This research was partly conducted for the ZIM SISSeC Project under Contract no. 16KN085725 from the German Federal Ministry of Economic Affairs and Energy.

References

[1] S. Boschert, C. Heinrich and R. Rosen, Next generation digital twin, *Proceedings of the Twelfth International Symposium on Tools and Methods of Competitive Engineering*, pp. 209–217, 2018.

[2] F. Breitinger and H. Baier, Similarity preserving hashing: Eligible properties and a new algorithm MRSH-v2, *Proceedings of the Fourth International Conference on Digital Forensics and Cyber Crime*, pp. 167–182, 2012.

[3] C. Chan, K. Chow, S. Yiu and K. Yau, Enhancing the security and forensic capabilities of programmable logic controllers, in *Advances in Digital Forensics XIV*, G. Peterson and S. Shenoi (Eds.), Springer, Cham, Switzerland, pp. 351–367, 2018.

[4] M. Dietz and G. Pernul, Digital twins: Empowering enterprises towards a system-of-systems approach, *Business and Information Systems Engineering*, vol. 62(2), pp. 179–184, 2020.

[5] M. Dietz and G. Pernul, Unleashing the digital twin's potential for ICS security, *IEEE Security and Privacy*, vol. 18(4), pp. 20–27, 2020.

[6] `dugpit`, `cdpfgl`: Continuous Data Protection for GNU/Linux, GitHub (`github.com/dupgit/sauvegarde`), 2021.

[7] M. Eckhart and A. Ekelhart, A specification-based state replication approach for digital twins, *Proceedings of the Workshop on Cyber-Physical Systems Security and Privacy*, pp. 36–47, 2018.

[8] M. Eckhart and A. Ekelhart, Towards security-aware virtual environments for digital twins, *Proceedings of the Fourth ACM Workshop on Cyber-Physical System Security*, pp. 61–72, 2018.

[9] M. Eckhart and A. Ekelhart, Digital twins for cyber-physical systems security: State of the art and outlook, in *Security and Quality in Cyber-Physical Systems Engineering*, S. Biffl, M. Eckhart, A. Lüder and E. Weippl (Eds.), Springer, Cham, Switzerland, pp. 383–412, 2019.

[10] M. Eckhart, A. Ekelhart and E. Weippl, Enhancing cyber situational awareness for cyber-physical systems through digital twins, *Proceedings of the Twenty-Fourth IEEE International Conference on Emerging Technologies and Factory Automation*, pp. 1222–1225, 2019.

[11] L. Englbrecht, DTDFEvaluation, GitHub (`github.com/Ludwig Englbrecht/DTDFEvaluation`), 2021.

[12] L. Englbrecht, `sauvegardeEX`, GitHub (`github.com/LudwigEngl brecht/sauvegardeEX`), 2021.

[13] L. Englbrecht, S. Meier and G. Pernul, Towards a capability maturity model for digital forensic readiness, *Wireless Networks*, vol. 26(7), pp. 4895–4907, 2020.

[14] M. Farsi, A. Daneshkhah, A. Hosseinian-Far and H. Jahankhani (Eds.), *Digital Twin Technologies and Smart Cities*, Springer, Cham, Switzerland, 2020.

[15] C. Gehrmann and M. Gunnarsson, A digital twin based industrial automation and control system security architecture, *IEEE Transactions on Industrial Informatics*, vol. 16(1), pp. 669–680, 2020.

[16] M. Grieves and J. Vickers, Digital twin: Mitigating unpredictable, undesirable emergent behavior in complex systems, in *Transdisciplinary Perspectives on Complex Systems*, F. Kahlen, S. Flumerfelt and A. Alves (Eds.), Springer, Cham, Switzerland, pp. 85–113, 2017.

[17] International Electrotechnical Commission, IEC 61131-3:2013 Programmable Controllers – Part 3: Programming Languages, Geneva, Switzerland, 2013.

[18] K. Kent, S. Chevalier, T. Grance and H. Dang, Guide to Integrating Forensic Techniques into Incident Response, NIST Special Publication 800-86, National Institute of Standards and Technology, Gaithersburg, Maryland, 2006.

[19] P. Kieseberg and E. Weippl, Security challenges in cyber-physical production systems, in *Software Quality: Methods and Tools for Better Software and Systems*, D. Winkler, S. Biffl and J. Bergsmann (Eds.), Springer, Cham, Switzerland, pp. 3–16, 2018.

[20] W. Kritzinger, M. Karner, G. Traar, J. Henjes and W. Sihn, Digital twins in manufacturing: A categorical literature review and classification, *IFAC-PapersOnLine*, vol. 51(11), pp. 1016–1022, 2018.

[21] Y. Liu, L. Zhang, Y. Yang, L. Zhou, L. Ren, F. Wang, R. Liu, Z. Pang and M. Deen, A novel cloud-based framework for elderly healthcare services using digital twins, *IEEE Access*, vol. 7, pp. 49088–49101, 2019.

[22] M. Lu and T. Chiueh, File versioning for block-level continuous data protection, *Proceedings of the Twenty-Ninth IEEE International Conference on Distributed Computing Systems*, pp. 327–334, 2009.

[23] E. Negri, L. Fumagalli and M. Macchi, A review of the roles of digital twins in CPS-based production systems, in *Value Based and Intelligent Asset Management: Mastering the Asset Management Transformation in Industrial Plants and Infrastructures*, A. Crespo Marquez, M. Macchi and A. Parlikad (Eds.), Springer, Cham, Switzerland, pp. 291–307, 2020.

[24] G. Palmer, A Road Map for Digital Forensic Research, DFRWS Technical Report, DTR-T001-01 Final, Air Force Research Laboratory, Rome, New York, 2001.

[25] J. Rubio, R. Roman and J. Lopez, Analysis of cybersecurity threats in Industry 4.0: The case of intrusion detection, *Proceedings of the International Conference on Critical Information Infrastructures Security*, pp. 119–130, 2017.

[26] G. Schroeder, C. Steinmetz, C. Pereira and D. Espindola, Digital twin data modeling with automationML and a communication methodology for data exchange, *IFAC-PapersOnLine*, vol. 49(30), pp. 12–17, 2016.

[27] F. Servida and E. Casey, IoT forensic challenges and opportunities for digital traces, *Digital Investigation*, vol. 28(S), pp. S22–S29, 2019.

[28] Y. Sheng, D. Wang, J. He and D. Ju, TH-CDP: An efficient block level continuous data protection system, *Proceedings of the International Conference on Networking, Architecture and Storage*, pp. 395–404, 2009.

[29] shramos, Polymorph (v2.0.5), GitHub (github.com/shramos/poly morph), 2020.

[30] J. Tan, Forensic readiness: Strategic thinking on incident response, presented at the *Second Annual CanSecWest Conference*, 2001.

[31] F. Tao, J. Cheng, Q. Qi, M. Zhang, H. Zhang and F. Sui, Digital twin driven product design, manufacturing and service with big data, *International Journal of Advanced Manufacturing Technology*, vol. 94(9), pp. 3563–3576, 2018.

[32] T. Wu, F. Breitinger and S. O'Shaughnessy, Digital forensic tools: Recent advances and enhancing the status quo, *Digital Investigation*, vol. 34, article no. 300999, 2020.

[33] K. Yau, K. Chow and S. Yiu, A forensic logging system for Siemens programmable logic controllers, in *Advances in Digital Forensics XIV*, G. Peterson and S. Shenoi (Eds.), Springer, Cham, Switzerland, pp. 331–349, 2018.

[34] X. Yu, Y. Tan, Z. Sun, J. Liu, C. Liang and Q. Zhang, A fault-tolerant and energy-efficient continuous data protection system, *Journal of Ambient Intelligence and Humanized Computing*, vol. 10(8), pp. 2945–2954, 2019.

Chapter 3

COMPARISON OF CYBER ATTACKS ON SERVICES IN THE CLEARNET AND DARKNET

York Yannikos, Quang Anh Dang and Martin Steinebach

Abstract   Cyber attacks on clearnet services are discussed widely in the research literature. However, a systematic comparison of cyber attacks on clearnet and darknet services has not been performed. This chapter describes an approach for setting up and simultaneously running honeypots with vulnerable services in the clearnet and darknet to collect information about attacks and attacker behavior. Key observations are provided and the similarities and differences regarding attacks and attacker behavior are discussed.

Keywords: Clearnet, darknet, services, honeypots, attacks, attacker behavior

1.   Introduction

Almost every publicly-available Internet service is subjected to cyber attacks. Much research has focused on attack frequencies, attack patterns and attacker behavior. Research has also investigated attacks on services in the clearnet by deploying honeypots [13, 25]. However, very little is known about attacks on services in the darknet, including onion services in the Tor network. At this time, there are no insights into how often clearnet services face cyber attacks compared with darknet services, the sophistication of the attacks and the motives of the attackers. Another open question is how automated attacks on clearnet services compare with those on darknet services.

Considerable research has focused on malware collection, traffic analysis and honeypots in the darknet [2, 11, 18, 21, 22]. However, the research essentially considers the darknet as a network telescope – public IP addresses that are not assigned to legitimate hosts [16] – instead

© IFIP International Federation for Information Processing 2021
Published by Springer Nature Switzerland AG 2021
G. Peterson and S. Shenoi (Eds.): Advances in Digital Forensics XVII, IFIP AICT 612, pp. 39–61, 2021.
https://doi.org/10.1007/978-3-030-88381-2_3

of an overlay network with a strong focus on anonymity as in the case of the Tor network.

This chapter discusses the similarities and differences between attacks observed in the clearnet and darknet using honeypots. The approach towards implementing, deploying and observing multiple honeypots in the two environments during the same time span is also detailed. Diverse attack vectors were considered by covering multiple protocols such as HTTP, SSH and Telnet that are attractive attack targets. The collected data was analyzed with a focus on automated attacks, attack frequencies and attacker behavior.

2. Background

Several definitions have been proposed for the clearnet and darknet. This section clarifies the definitions based on recent work by Kavallieros et al. [14]:

- **World Wide Web:** The World Wide Web, or simply the web, is an important part of the Internet. It houses a massive collection of diverse documents, many of which are linked. The documents are usually accessed via web browsers using the HTTP or HTTPS protocols. Other important components of the Internet, which are not part of the web, include services that support file transfer, email and hostname resolution.

- **Surface Web:** The surface web, also called the visible web, is a subset of the web that holds all the indexed web content on the Internet. The indexed web content is accessed using search engines like Google and Bing. Of course, the surface web only holds a small portion of the total content available on the Internet.

- **Deep Web:** The deep web is a subset of the web holding content that is not indexed by search engines. The reasons for content not being indexed vary. Examples are private web resources with no external links pointing to them and web content accessible only via login credentials or a virtual private network. Because search engines continuously expand and refine their indexing capabilities, the deep web and surface web content are always in flux.

- **Darknet:** The darknet is a subset of the Internet that comprises many separated networks. These decentralized networks function as overlay networks on top of the Internet infrastructure. The networks, which are only accessed via additional software, support anonymous participation and communications. The darknet can provide content from the deep web and surface web as well as non-web services.

- **Dark Web:** The dark web is essentially the web portion of the darknet. It holds all the web content of the darknet. The content is usually accessed via web browsers using the HTTP or HTTPS protocols.

- **Clearnet:** The clearnet is the counterpart to the darknet. Every service, participant and content in the Internet that is not in the darknet is part of the clearnet. The union of the clearnet and darknet is the Internet.

This research uses the terms clearnet and darknet as defined above. The Tor network in the darknet was chosen as the testing environment due to its popularity:

- **Tor Network:** Tor, which stands for The Onion Router, is one of the overlay networks of the darknet. Tor provides anonymity to its users via onion routing, an ingenious way of routing packets that also leverages several layers of encryption. Every packet in the Tor network is routed via a predefined path with a minimum of three nodes, ensuring that the source node of a packet is never directly connected to its destination node. The routing paths are called Tor circuits.

  Tor supports two important features. It enables its participants to anonymously access clearnet services outside Tor as well as anonymously access host-specific services called onion services in Tor. Onion services typically host websites, but almost any other common Internet service can be hosted (e.g., services that support file transfer, email and chats).

  Since Tor is the most popular darknet network, it is often used as a synonym for the entire darknet.

3.   Common Targets and Attacks

This section describes the services that are commonly targeted by cyber attacks. Based on resources such as OWASP Top Ten [20], this section also discusses the types of attacks expected when deploying honeypots hosting commonly-targeted services.

Common targets of cyber attacks are:

- **Web Servers:** As in the clearnet, web content is the most-accessed content in the darknet. Therefore, web servers are attractive targets for attackers, especially those seeking quick monetary gains (e.g., taking over Tor marketplaces or bitcoin escrow services).

- **Remote Access Services:** Remote access services such as SSH are typically used for system administration and are regularly attacked. In the worst case, a vulnerable SSH server, insecure configuration or weak login credentials for a root account can enable an attacker to take over a system that hosts a variety of services.

- **Email Servers:** Email servers are popular targets for attackers. Successful attacks often provide access to sensitive information contained in email. Additionally, email servers can be leveraged to launch other attacks such as spamming and phishing.

- **Database Servers:** Database servers are targeted for the sensitive information maintained in databases. The servers should not be directly accessible from the Internet, but this is often not the case. Several prominent data leaks are the direct result of targeting insecure, remotely-accessible database servers.

Attackers typically employ reconnaissance techniques such as port scanning and fingerprinting to gather information about potential targets. Tools such as Nmap, ZMap and Masscan support efficient port scanning of large portions of the Internet. Attacks are launched based on the open ports that are discovered and the services they support. Common attacks are:

- **Injection Attacks:** SQL, NoSQL, LDAP and other injection attacks are very common on the Internet [20]. Injection attacks are often successful when untrusted user inputs, such as SQL database queries, are not properly checked and sanitized before they are processed. Services vulnerable to injection attacks can be found and exploited automatically using tools such as `sqlmap` and Metasploit.

- **Brute Force Attacks:** Brute force attacks are commonly used against services that require authentication with login credentials. Examples are websites, SSH and FTP. Brute force attacks can be automated very easily when two requirements are met. First, the server's response to a failed login attempt is distinguishable from a successful attempt (which is almost always the case). Second, the server does not use a mechanism such as reCAPTCHA to defeat automated requests.

- **Cross-Site Scripting Attacks:** Cross-site scripting (XSS) attacks are included in the OSWASP list of common attacks against web servers. A cross-site scripting attack is an injection attack where untrusted user input is (temporarily or permanently) embedded in a web page. If the input is not properly sanitized, then an attacker could, for example, embed JavaScript code in a web page that is executed when a victim visits the page. The embedded code could steal cookies, hijack a user session and/or redirect the victim to a malicious website.

4.   Related Work

The darknet and its offerings have drawn the attention of many researchers in recent years. This section provides an overview of research related to the Tor network.

Catakoglu et al. [6] conducted research similar to that described in this chapter. They deployed a honeypot service to collect and analyze data pertaining to attacks in the darknet. In addition to a web service, they deployed other services such as SSH and IRC in their honeypots. However, they did not compare simultaneously-running services in the darknet and clearnet to gain insights into the similarities and differences in attacker behavior.

Zeid et al. [26] used two honeypots to collect evidence of malicious and illegal activities in the darknet. They installed a chatroom and a vulnerable website as onion services in Tor, and collected information about individuals who searched for child pornography and hacking techniques. They also developed an advanced chatbot for the darknet by training a recurrent neural network with data collected from two large darknet marketplaces [17].

From 2013 through 2015, Branwen [3] scraped a large amount of data from 89 Tor marketplaces and made it publicly available. Several researchers have analyzed the data. For example, Broseus et al. [5] conducted a comprehensive analysis of data about global trafficking in darknet marketplaces, providing valuable insights into the types of goods, originating countries and specializations of the dealers.

Steinebach et al. [23] analyzed the popularity of hidden services and demonstrated that a significant percentage of the accessed darknet services belonged to botnets. Other popular services included drug markets, hacking forums and websites with political content. Flamand and Decary-Hetu [12] obtained insights into the human factors underlying cyber crime by analyzing drug offerings in the Tor network with a focus on the entrepreneurship activities of online drug dealers.

5.   Honeypot Deployment

A three-phase process was applied to deploy honeypot services on two virtual machines (VMs), one each in the clearnet and darknet. This section describes the security considerations and the phases of the deployments.

5.1   Security Considerations

Because the goal was to have the honeypot services attacked, security issues pertaining to the virtual machines had to be minimized. These included the complete takeover of the virtual machines, deletion of virtual machine data and using the virtual machines to launch internal and external attacks. Therefore, the virtual machines were placed in an isolated network protected by strict firewall rules. Additionally, each honeypot service was executed in an isolated environment in the virtual machines. Details about the firewall configuration and service isolation are provided below.

5.2   Deployment Process

The deployment of honeypot services involved three phases:

- **Phase 1:** In the first phase, two popular services – a web service and an SSH service – with vulnerabilities were deployed on the clearnet and darknet virtual machines. The two services constituted the foundation of the honeypots. In order to announce the availability of the honeypot services on the Internet, a domain name was registered and an onion address was created for the clearnet and darknet virtual machines, respectively. All activities involving the deployed honeypot services were monitored. After a few days of monitoring, the availability of the services was announced using various platforms and search engines. The services were not announced immediately because the intent was to first check if traffic related to automated scans for web or SSH services was observed. Although it was very unlikely that the onion address of the darknet virtual machine would be found by a visitor purely by guessing, it was expected that the clearnet virtual machine would encounter some web or SSH traffic before the announcement.

- **Phase 2:** In the second phase, all the ports were monitored, and the ports (and HTTP paths) that were most frequently scanned were analyzed. Based on the number of observed port scans, it was decided to deploy additional honeypot services for Telnet, SMTP and FTP. In the case of the darknet virtual machine, high interest in port 443 for HTTPS was observed. Therefore, an HTTPS proxy with a self-signed certificate was deployed on the darknet virtual machine.

- **Phase 3:** In the third phase, adjustments were made to the honeypot services based on the numbers of HTTP path scans and attacks observed during the second phase. To increase the probability of receiving attack traffic, the honeypot services were announced on additional platforms (e.g., by posting links on social networks). Additional vulnerabilities were also added to the honeypot services to expand the attack surface.

6. Implementation Details

This section describes the virtual machine architectures and the honeypot service implementations.

6.1 Virtual Machine Architectures

Figures 1 and 2 present the architectures of the virtual machines deployed to observe attackers in the clearnet and darknet, respectively. The virtual machines used the Arch Linux operating system and docker to run each honeypot service in an isolated manner. An additional SSH service for administrative purposes and a MariaDB instance for the central storage of logged data were installed on each virtual machine. Also, iptables was installed on each virtual machine to serve as a firewall, specifically to control the amount and type of traffic reaching each honeypot service. The iptables configurations forwarded traffic to the services and simultaneously logged the traffic on all ports. To prevent outgoing attacks, any and all attempts to create new connections from the virtual machines to external systems were blocked.

A Tor proxy was installed on the darknet virtual machine because it required a Tor connection. All the IP addresses that requested honeypot services on the clearnet virtual machine were monitored; however, it was not possible to collect IP addresses when using Tor. Therefore, the Tor proxy was configured to log the IDs of the Tor circuits associated with requests and prevent early closures of the circuits. This enabled the mapping of requests via Tor to the Tor circuits used by individuals.

6.2 Honeypot Services

The honeypot services were deployed in separate docker containers:

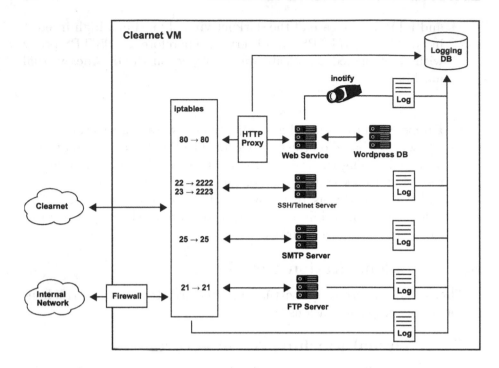

Figure 1.   Architecture of the clearnet virtual machine.

- **Web:** Web services were deployed in two docker containers, one
  with an Apache web server (version 2.4.25 from December 2016)
  hosting a Wordpress instance and the other with the MySQL server
  required for Wordpress. Wordpress was used as an image gallery
  with pictures of cats and dogs. Harmless content was chosen over
  political, religious and other controversial content because the intent was not to attract attackers with specific agendas. Instead,
  the goal was to attract attackers based on the technical vulnerabilities of the deployed software.

  A small HTTP proxy was installed in front of the Apache web
  server to filter certain HTTP requests, perform geolocation lookups
  of IP addresses (clearnet virtual machine only), and log all HTTP
  requests (including payloads) and responses. In the case of the
  darknet virtual machine, an HTTPS proxy using a self-signed certificate was also installed.

  Two versions of honeypot web services were employed. The first
  version, which employed Wordpress version 4.9 from November
  2017, incorporated the following additional vulnerabilities:

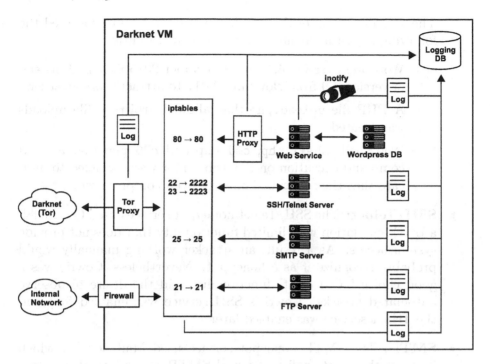

Figure 2.   Architecture of the darknet virtual machine.

- Verbose directory listings to disclose files in directories with no index pages.

- Multiple custom login and registration forms with weak credentials to gain database access. The credentials could also be used for authentication by the SSH/Telnet services.

- A built-in SQL injection vulnerability to bypass authentication using the login form.

- A file providing **phpinfo()** and other verbose PHP errors to support information disclosure.

- A customized 404 error page in PHP with a built-in cross-site scripting vulnerability.

A crawler trap was set up to distinguish automated attacks from manual attacks. Specifically, an empty HTML page with a random (i.e., hard to guess) name was inserted in the root directory and a link to it was placed in the index page using a small (10×1 pixel) invisible image. A **robots.txt** file that referred to files suggestive of sensitive content was also created.

The second version of the web honeypot service incorporated the following adjustments and additional vulnerabilities:

- Wordpress version 4.4 from December 2015 was used instead of version 4.9 from November 2017 to attract more attackers.
- A PHP file `upload.php` that allowed arbitrary file uploads was created.
- Web shells `shell.php`, `cmd.php` and `c99.php` that allowed command execution on the web server were installed to indicate that the server had already been compromised.

- **SSH/Telnet:** The SSH/Telnet honeypot services used Cowrie [8], a UNIX emulation with limited functionality that does not provide system access. As a result, an attacker working manually would probably recognize it as a honeypot. Nevertheless, Cowrie was a good choice based on its functionality and the desire to observe automated attacks. Cowrie's SSH service was enabled first and the Telnet service was enabled later.

- **SMTP:** The SMTP honeypot service used Mailoney [1], which imitates the functionality of a real SMTP server but silently logs and discards all sent emails. It was configured to enable email to be sent without authentication, thereby imitating an open mail relay.

- **FTP:** An FTP honeypot [4] written in Python was installed to masquerade as a real FTP server. The FTP service was configured to allow anonymous and authenticated uploads with weak credentials and log all FTP commands.

7.   Experiments and Results

This section describes the steps involved in deploying and announcing the honeypot services hosted by the virtual machines in the clearnet and darknet. Data was collected from December 29, 2018 through March 14, 2019. This section also presents the results obtained by analyzing the attacker traffic logged on the virtual machines.

7.1   Service Deployments

Web and SSH honeypot services were deployed on December 29, 2018. Additional honeypot services were deployed and existing honeypot services were adjusted incrementally to attract attackers. Table 1 shows the deployment dates and uptimes of the honeypot services.

Table 1. Chronological list of honeypot service deployments.

Service	Start	End	Uptime
Web honeypot (v1)	29.12.2018	22.02.2019	56d
SSH honeypot	29.12.2018	14.03.2019	76d
iptables	14.01.2019	14.03.2019	60d
Telnet honeypot	20.01.2019	14.03.2019	54d
SMTP honeypot	24.01.2019	14.03.2019	50d
FTP honeypot (port 2121)	24.01.2019	25.02.2019	33d
HTTPS proxy (darknet)	13.02.2019	14.03.2019	26d
Web honeypot (v2)	23.02.2019	14.03.2019	20d
FTP honeypot (port 21)	26.02.2019	14.03.2019	17d

7.2   Announcements

After deploying the virtual machines during the first phase, the honeypot services were announced on the Internet a few days later. The announcements, which included web addresses or onion addresses, were issued as posts or comments on platforms such as Reddit. Indexing requests were also sent to several search engines. Table 2 shows the dates of the announcements issued after the initial deployment.

7.3   Observed Web Requests

It was necessary to distinguish automated web requests from manual requests. The following criteria were used to identify web requests as automated (bot) traffic:

- Crawler trap, `robots.txt` and referenced files were accessed.

- Common IP address ranges for bots from Google, Bing and other search engines were employed (clearnet only).

- The IP address instead of the domain name was targeted (clearnet only).

- Non-existent files or directories not linked to or visible on the website were requested (e.g., using automated web path scanners).

- User agents of common scanners, bots or similar tools were employed instead of typical web browsers.

After 56 days of monitoring the first version of the web honeypot service, no attempts to exploit the custom SQL injection or cross-site scripting vulnerabilities were observed. However, activities by several

Table 2.   Chronological list of web and onion address announcements on the Internet.

Date	Event
29.12.2018	Initial deployment of web and SSH honeypot services in the clearnet and darknet.
14.01.2019	Announcement of onion addresses in several subforums of Reddit ("subreddits") and the 4chan bulletin board, announcement of web addresses on 4chan, sending of web address indexing requests to Google, sending of onion address indexing requests to several Tor search engines.
17.01.2019	Announcement of web addresses as comments on several YouTube videos and subreddits related to cats.
25.01.2019	Announcement of web addresses on different subreddits.
10.02.2019	Posting of the onion addresses on Hidden Wiki and Onion List.
26.02.2019	Submission of web addresses to different subreddits and other link aggregators, announcement of Tor2Web proxy URLs.

tools that scanned web paths were observed, especially requests for common web shells. Based on these observations, the second version of the web honeypot service with additional vulnerabilities (namely, arbitrary file uploads and web shells) was deployed. The second version was monitored for 20 days.

Clearnet.   The web honeypot service on the clearnet virtual machine fielded 32,605 HTTP requests (69.6% GET, 30.1% POST, 0.1% HEAD) in 4,157 sessions (i.e., requests from the same IP address within a short time span) during the 76 days of uptime. A total of 2,931 unique IP addresses originating from 113 countries were logged. Application of the criteria listed above classified 79.7% of the sessions as automated (bot) sessions; 67.3% of the web requests were issued in these sessions. Figure 3 shows the distribution of web requests by country.

The logs recorded 7,918 GET requests (24.3% of the total) that were used for web path scanning (i.e., attempting to find hidden files via brute force). Examples of the scanned paths were:

- Database administration software (e.g., `/phpmyadmin`).
- A file used by the Muhstik botnet [19] (i.e., `muhstik.php`).
- WebDAV directories (e.g., `/webdav`).
- VoIP administration software (e.g., `/voip` and `/cisco`).

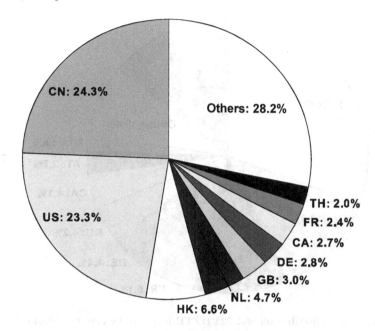

Figure 3. Distribution of 32,605 HTTP requests by country (clearnet).

The logs recorded 6,906 POST requests (21.2% of the total) that were attempting to launch a PHP code injection attack against a non-existent file /test.php. The requests originated from 63 IP addresses in 12 countries (10 countries in Asia, the majority from China, Hong Kong and Taiwan). Since all the attacks shared the same pattern, they were likely launched from a botnet in the region.

The logs also recorded 2,873 POST requests (8.8% of the total) involved in brute force attacks against the Wordpress instance. The requests originated from 701 IP addresses in 54 countries. All the requests used the same user agent, an indication that the brute force attacks were launched from a single botnet.

Other attacks included small numbers of attempts to exploit vulnerabilities in D-Link modems [9], ThinkPHP [10] and Avtech IP cameras [7]. Just before the monitoring of the web honeypot service ended, an attacker from an IP address in the United Kingdom accessed the /shell.php web shell and manually attempted to gather information. The attacker eventually issued a command to wipe all the system data.

Darknet. The web honeypot service on the darknet virtual machine fielded 112,082 HTTP requests (97.6% GET, 0.1% POST, 2.2% HEAD)

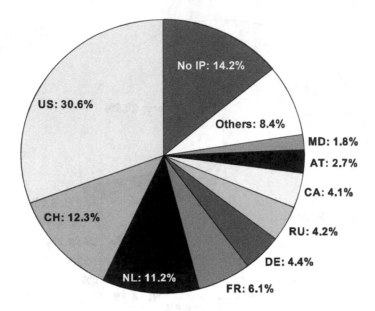

Figure 4.   Distribution of 1,728 HTTP requests by country (darknet).

in 7,784 sessions during 76 days of uptime. Of these requests, 24.4% were classified as automated (bot) and 75.6% as manual requests.

The logs recorded 1,728 (1.5%) HTTP requests sent using a Tor2Web proxy. Upon examination, 85.8% of these requests contained HTTP headers that revealed the originating IP address (i.e., headers such as `X-Real-Ip`, `X-Tor2web` and `X-Forwarded-Host`). Most of the requests came from the United States and European countries. Figure 4 shows the distribution of web requests issued using a Tor2Web proxy by country.

The logs recorded 14,879 GET requests (13.3% of the total) that were used for web path scanning. Examples of the scanned paths were:

- Database dump files in various formats (e.g., `mysql.zip`, `dump.sql` and `dump.sql.zip`).

- Private key file of the onion service that can be used to steal the onion domain name (i.e., `/private_key`).

- Files related to cryptocurrencies (e.g., `wallet.zip`, `wallet.dat` and `bitcoin.php`).

- Files and path traversal attempts to gather information about Linux systems (e.g., `/etc`, `/.ssh` and `.bash_history`).

- Web shells (e.g., `/shell.php`).

- WebDAV directories (e.g., `/webdav`).

No attacks using GET requests were logged during the uptime. Several registration and login attempts using the custom forms were attempted, but no attempts to exploit the custom SQL injection or cross-site scripting vulnerabilities were observed.

Upon deploying the second version of the web honeypot server, a manual attack that lasted more than 17 minutes was observed. The attacker located the arbitrary file upload script `upload.php` and used it to upload the `wow.php` web shell. A total of 19 POST requests addressing `wow.php` were observed, each with a different user agent. Finally, the attacker uploaded the `c99.php` shell and modified the permissions of `/var/www/html`, which resulted in the web page becoming inaccessible.

Comparison. The darknet virtual machine had 3.4 times as many HTTP requests as the clearnet virtual machine. However, a much larger proportion of automated traffic/attacks (67.3% of all the HTTP requests) was observed on the clearnet virtual machine compared with the darknet virtual machine (24.4%). The web honeypot service on the darknet virtual machine had more visits likely out of curiosity after the website was announced. In absolute terms, both the web services received similar numbers of automated HTTP requests, about 22K for the clearnet and 27K for the darknet honeypots.

Visiting bots from major search engines such as Google were observed on both web services. The web service on the darknet virtual machine was crawled via a Tor2Web proxy, resulting in both web services being indexed (e.g., by Google). Also, both the web services experienced automated brute force attacks against the installed Wordpress instances, directly targeting specific pages such as `/xmlrpc.php` and `/wp-login.php`. This may indicate that automated vulnerability scanners use search engines to look for vulnerable Wordpress installations in the clearnet as well as the darknet. Path scanners appeared to perform different searches in the clearnet and darknet (e.g., looking for data related to cryptocurrencies), but similar interests were observed with respect to database dumps, administration tools and web shells.

Most attackers addressed the web honeypot service on the clearnet virtual machine using its IP address instead of its domain name. This may be another indication that automated tools were used to scan and attack large IP address ranges. In the case of the clearnet virtual machine, many attacks attempted to target popular software vulnerabilities directly without first checking if the software was actually installed. In the case of the darknet virtual machine, more initial reconnaissance traffic was observed along with interest in darknet-specific data such as the onion service private key.

7.4   Observed SSH and Telnet Access

This section presents details about and comparisons of the SSH and Telnet honeypot service access on the clearnet and darknet virtual machines.

Clearnet.   The clearnet virtual machine had incoming traffic to the SSH honeypot just 62 minutes after the initial deployment, even before any announcements were made. During the uptime, a total of 1,194,952 SSH and Telnet sessions (85.6% SSH and 14.4% Telnet) were logged; they originated from 18,345 IP addresses located in 177 countries. About 77.6% of the sessions achieved successful logins by targeting the weak login credentials. About 92% of the login attempts sought to brute force the root user password. The most popular password attempted was root (18.8%).

The logs recorded about 890K sessions during which attackers attempted to utilize the SSH honeypot as a SOCKS proxy for tunneling their traffic (e.g., to attack other targets). The most popular targets for access using SSH were web (HTTP(S)) and email (SMTP/IMAP/POP3) services. Table 3 lists the most popular targets along with the numbers of SSH sessions during which connections to the targets were attempted using the honeypot as a SOCKS proxy. The majority of the HTTP requests were simple GET requests without payloads or query parameters. These requests were likely issued to check if the SOCKS proxy was working as intended.

Attacker activity on the SSH honeypot was analyzed by examining the commands issued in the SSH sessions. From among the approximately 900K shell commands (10,018 unique commands) issued in 83K sessions, a large number of sessions (33,732 corresponding to 40.5% of all the sessions) had activity patterns associated with the Mirai botnet [15]. The commands fell into six categories:

- **Mirai:** Typical commands known to be issued by the Mirai botnet.

- **Info:** Commands used to gather information (e.g., ping, whoami, uptime, cat and ls).

- **Harmless:** Commands with no significant effects (e.g., exit) and empty commands.

- **Script:** Nested and scripted commands (e.g., multiple actions in a command line).

Table 3. Service connections attempted using the SSH honeypot.

Domain Name/IP Address	Protocol	Sessions
ya.ru	HTTP	301,208
163.172.20.152	HTTP	39,566
bot.whatismyipaddress.com	HTTP	16,643
2a02:6b8:a::a 6574	HTTP	3,249
186.190.212.26	HTTP	2,984
ya.ru	HTTPS	478,923
www.google.com	HTTPS	28,197
www.walmart.com	HTTPS	10,006
www.netflix.com	HTTPS	8,970
www.google.co.uk	HTTPS	7,856
96.114.157.80	SMTP	7,215
68.87.20.5	SMTP	7,201
98.136.101.117	SMTP	7,145
66.218.85.52	SMTP	6,870
67.195.229.59	SMTP	6,870
imap.apple.mail.yahoo.com	IMAP/POP3	11,534
imap.email.comcast.net	IMAP/POP3	4,264
imap.aol.com	IMAP/POP3	4,192
imap.gmx.net	IMAP/POP3	508
imap-mail.outlook.com	IMAP/POP3	482

- **Change:** Commands used to change filesystem permissions, download files, free memory, etc.

- **Other:** Commands not belonging to the five preceding categories.

Table 4. Commands per category issued to the SSH/Telnet honeypot.

Category	Commands	SSH	Telnet	IP Addresses	Countries
Mirai	746,871	20	33,712	2,774	119
Info	140,616	2,104	33,887	4,669	138
Harmless	13,719	26	13,363	71	16
Script	129	67	61	86	25
Change	50	30	1	13	10
Other	32	23	1	6	4

Table 4 shows the numbers of commands in the six categories issued by attackers on the SSH/Telnet honeypot. Associated with each command category are the numbers of SSH and Telnet sessions and originating IP addresses and countries.

Telnet appears to be a popular service for issuing Mirai botnet and information gathering commands. In 128 sessions (0.2%), attackers used similar numbers of fully-scripted commands on SSH and Telnet (e.g., to download and deploy backdoors or remove traces from log files).

Examination of the files downloaded by the attackers revealed multiple samples that were presumably intended or used to launch large-scale automated attacks. For example, six shell scripts were found to share the same pattern: they first tried to download and then run 10 different executable files of the same program, each compiled for a different processor architecture (e.g., x68, MIPS, PowerPC and ARM).

Darknet. Although the SSH service on the clearnet virtual machine had a lot of activity, the SSH service on the darknet virtual machine received very little attention. In total, 184 sessions (50% SSH and 50% Telnet) were logged and only after the onion address was published for the first time on January 14, 2019. Compared with the clearnet virtual machine, no login attempts were observed on the darknet virtual machine. All 184 sessions appeared to originate from port scanners or banner grabbers.

Comparison. Compared with the darknet virtual machine, almost 6,500 times as many SSH and Telnet sessions were logged on the clearnet virtual machine. While the SSH/Telnet honeypot was subjected to many different attacks such as brute force login attempts, use as a SOCKS proxy and automated botnet attacks, the honeypot on the darknet virtual machine was not attacked at all and received very little reconnaissance traffic.

7.5   Observed SMTP Requests

This section presents details about and comparisons of the SMTP honeypot service access on the clearnet and darknet virtual machines.

Clearnet. During the uptime of the SMTP honeypot, 2,502 sessions were logged on the clearnet virtual machine; they originated from 130 IP addresses in 27 countries. Approximately 90% of the sessions originated from Russia. A closer look of the logged data revealed that they came from the PushDo/Cutwail botnet [24] that attempted to send spam email. However, since the attackers only sent a few packets in most SMTP sessions, only 74 attacker email messages were captured. Of the

74 captured email messages, 46 (62.2%) were empty messages or test messages, the other 28 (37.8%) were phishing email messages.

Darknet. Absolutely no traffic was observed on the SMTP honeypot installed on the darknet virtual machine.

7.6   Observed FTP Requests

This section presents details about and comparisons of the FTP honeypot service access on the clearnet and darknet virtual machines.

Clearnet. The FTP honeypot installed on the clearnet virtual machine had 205 FTP sessions that originated from 131 IP addresses in 24 countries. One hundred of the 205 sessions originated from 76 IP addresses in the United States. Across all sessions, the following activities were observed:

- A total of 86 banner grabbing attempts.

- A total of 16 HTTP GET requests, including four requests issued from IPIP.NET, presumably for research purposes.

- Four attempts to gather information via FTP commands without logging in (e.g., HELP, STAT and LIST).

- A total of 99 login attempts,

Of the 99 login attempts, 41 were successful – 35 as **anonymous** and six as **root**. Of these, 39 attempted to gain information using FTP commands such as CWD, FEAT, HELP, LIST and PWD. The remaining two authenticated sessions originated from the same IP address in Germany. The attacker logged in as **anonymous**, gathered information about the FTP server and downloaded a public-private key pair that was intentionally stored on the server. The attacker proceeded to create three directories and eventually uploaded a harmless file containing an unrelated OpenVPN traffic dump.

Darknet. Only 35 banner grabbing attempts from 31 unique circuits were logged by the FTP honeypot on the darknet virtual machine. No login attempts or attacks were observed.

Comparison. The clearnet virtual machine had almost six times the amount of FTP session activity seen on the darknet virtual machine. As in the case of the SSH/Telnet honeypot, the FTP honeypot hosted on the darknet did not receive any traffic apart from a few reconnaissance attempts.

7.7   Discussion

Analysis of the collected data revealed that services such as SMTP and
FTP are of much less interest to attackers than web, SSH and Telnet
services. In the darknet, only web services attracted attackers, which
supports the findings of Catakoglu et al. [6].

The majority of attacks on the honeypot services hosted in the clearnet were automated (e.g., from botnets). In contrast, the darknet web
service, which was the only darknet service to see any attacks, was targeted mostly via manual means. Additionally, in the darknet, a higher
amount of reconnaissance traffic preceded the attack attempts.

Certain similarities were observed between attacks on the clearnet and
darknet service deployments. The clearnet and darknet virtual machines
both encountered scans for web shells and database-related data as well
as brute force attempts that sought to exploit common Wordpress vulnerabilities. No attempts were made to exploit the SQL injection and
cross-site scripting vulnerabilities on the clearnet and darknet virtual
machines, neither by automated tools nor by manual means.

8.   Conclusions

This chapter has described an approach for setting up and simultaneously operating honeypots with vulnerable services to collect information
about attackers in the clearnet and darknet. The two systems, which
were deployed for 76 days, collected valuable information about attacks
conducted by individuals and bots.

Attacks common to the clearnet and darknet deployments focused
on Wordpress instances in the web honeypot services. While a large
number of automated (mostly botnet) attacks on web services were observed in the clearnet deployment, attacks on web services in the darknet
deployment were mostly conducted manually. Interestingly, custom vulnerabilities were neither discovered nor exploited in the clearnet and
darknet deployments. In the clearnet, attackers were more interested in
web, SSH and Telnet services than SMTP and FTP services. In contrast, only web services attracted attackers in the darknet deployment,
which confirms previous research results [6].

Future research will expand the range of the deployed honeypot services by installing additional vulnerable versions of popular content management systems. Additionally, the uptime for data collection will be
increased significantly.

Acknowledgement

This research was supported by the German Federal Ministry of Education and Research (BMBF) under the PANDA Project (`panda-pro jekt.de`).

References

[1] awhitehatter, Mailoney – An SMTP Honeypot, GitHub (`github.com/awhitehatter/mailoney`), 2021.

[2] E. Bou-Harb, M. Debbabi and C. Assi, A time series approach for inferring orchestrated probing campaigns by analyzing darknet traffic, *Proceedings of the Tenth International Conference on Availability, Reliability and Security*, pp. 180–185, 2015.

[3] G. Branwen, Darknet Market Archives (2013–2015) (`www.gwern.net/DNM-archives`), 2019.

[4] A. Bredo, `honeypot-ftp` – FTP Honeypot, GitHub (`github.com/alexbredo/honeypot-ftp`), 2014.

[5] J. Broseus, D. Rhumorbarbe, M. Morelato, L. Staehli and Q. Rossy, A geographical analysis of trafficking in a popular darknet market, *Forensic Science International*, vol. 277, pp. 88–102, 2017.

[6] O. Catakoglu, M. Balduzzi and D. Balzarotti, Attack landscape in the dark side of the web, *Proceedings of the Symposium on Applied Computing*, pp. 1739–1746, 2017.

[7] Cisco Systems, Announcement regarding non-Cisco product security alerts, San Jose, California (`tools.cisco.com/security/center/viewAlert.x?alertId=49625`), 2019.

[8] Cowrie, `cowrie` – Cowrie SSH/Telnet Honeypot, GitHub (`github.com/cowrie/cowrie`), 2021.

[9] Exploit Database, D-Link DSL-2750B – OS Command Injection (`www.exploit-db.com/exploits/44760`), May 25, 2018.

[10] Exploit Database, ThinkPHP 5.X – Remote Command Execution (`www.exploit-db.com/exploits/46150`), January 14, 2019.

[11] C. Fachkha and M. Debbabi, Darknet as a source of cyber intelligence: Survey, taxonomy and characterization, *IEEE Communications Surveys and Tutorials*, vol. 18(2), pp. 1197–1227, 2016.

[12] C. Flamand and D. Decary-Hetu, The open and dark web, in *The Human Factor of Cybercrime*, R. Leukfeldt and T. Holt (Eds.), Routledge, London, United Kingdom, pp. 34–50, 2019.

[13] D. Fraunholz, D. Krohmer, S. Anton and H. Schotten, Investigation of cyber crime conducted by abusing weak or default passwords with a medium interaction honeypot, *Proceedings of the International Conference on Cyber Security and Protection of Digital Services*, 2017.

[14] D. Kavallieros, D. Myttas, E. Kermitsis, E. Lissaris, G. Giataganas and E. Darra, Understanding the dark web, in *Dark Web Investigations*, B. Akhgar, M. Gercke, S. Vrochidis and H. Gibson (Eds.), Springer, Cham, Switzerland, pp. 3–26, 2021.

[15] C. Kolias, G. Kambourakis, A. Stavrou and J. Voas, DDoS in the IoT: Mirai and other botnets, *IEEE Computer*, vol. 50(7), pp. 80–84, 2017.

[16] D. Moore, C. Shannon, G. Voelker and S. Savage, Network Telescopes: Technical Report, Cooperative Association for Internet Data Analysis, San Diego Supercomputer Center, University of California San Diego, La Jolla, California, 2004.

[17] J. Moubarak and C. Bassil, On darknet honeybots, *Proceedings of the Fourth Cyber Security in Networking Conference*, 2020.

[18] K. Nakao, D. Inoue, M. Eto and K. Yoshioka, Practical correlation analysis between scan and malware profiles against zero-day attacks based on darknet monitoring, *IEICE Transactions on Information and Systems*, vol. E92-D(5), pp. 787–798, 2009.

[19] L. O'Donnell, Muhstik botnet exploits highly critical Drupal bug, *Threatpost*, April 23, 2018.

[20] OWASP Foundation, OWASP Top Ten, Bel Air, Maryland (`owasp.org/www-project-top-ten`), 2020.

[21] J. Song, J. Choi and S. Choi, A malware collection and analysis framework based on darknet traffic, in *Neural Information Processing*, T. Huang, Z. Zeng, C. Li and C. Leung (Eds.), Springer, Berlin Heidelberg, Germany, pp. 624–631, 2012.

[22] J. Song, J. Shimamura, M. Eto, D. Inoue and K. Nakao, Correlation analysis between spamming botnets and malware-infected hosts, *Proceedings of the International Symposium on Applications and the Internet*, pp. 372–375, 2011.

[23] M. Steinebach, M. Schafer, A. Karakuz, K. Brandl and Y. Yannikos, Detection and analysis of Tor onion services, *Proceedings of the Fourteenth International Conference on Availability, Reliability and Security*, article no. 66, 2019.

[24] B. Stone-Gross, T. Holz, G. Stringhini and G. Vigna, The underground economy of spam: A botmaster's perspective of coordinating large-scale spam campaigns, *Proceedings of the Fourth USENIX Conference on Large-Scale Exploits and Emergent Threats*, 2011.

[25] A. Vetterl and R. Clayton, Honware: A virtual honeypot framework for capturing CPE and IoT zero days, *Proceedings of the APWG Symposium on Electronic Crime Research*, 2019.

[26] R. Zeid, J. Moubarak and C. Bassil, Investigating the darknet, *Proceedings of the International Wireless Communications and Mobile Computing Conference*, pp. 727–732, 2020.

II

APPROXIMATE MATCHING TECHNIQUES

Chapter 4

USING PARALLEL DISTRIBUTED PROCESSING TO REDUCE THE COMPUTATIONAL TIME OF DIGITAL MEDIA SIMILARITY MEASURES

Myeong Lim and James Jones

Abstract   Digital forensic practitioners are constantly challenged to find the best allocation of their limited resources. While automation will continue to partially mitigate this problem, the preliminary question about which media should be prioritized for examination is largely unsolved. Previous research has developed methods for assessing digital media similarity that may aid in prioritization decisions. Similarity measures may also be used to establish links between media and, by extension, the individuals or organizations associated with the media. However, similarity measures have high computational costs that delay the identification of digital media warranting immediate attention and render link establishment across large collections of data impractical.

This chapter presents and validates a method for parallelizing the computations of digital media similarity measures to reduce the time requirements. The proposed method partitions digital media and distributes the computations across multiple processors. It then combines the results as an overall similarity measure that preserves the accuracy of the original method executed on a single processor. Experiments on a limited dataset demonstrate reductions of up to 51% in processing time. The reductions vary based on the number of partitions chosen and specific digital media being examined, suggesting the need for additional testing and optimization strategies.

Keywords: Drive similarity, sector hashes, Jaccard index, parallel computation

1.   Introduction

Digital forensic practitioners extract and process evidence from digital sources and media during the course of criminal and other investigations.

© IFIP International Federation for Information Processing 2021
Published by Springer Nature Switzerland AG 2021
G. Peterson and S. Shenoi (Eds.): Advances in Digital Forensics XVII, IFIP AICT 612, pp. 65–87, 2021.
https://doi.org/10.1007/978-3-030-88381-2_4

Digital evidence is fragile and volatile, requiring the attention of trained specialists to ensure that content of evidentiary value can be effectively isolated and extracted in a forensically-sound manner. This work is often time-sensitive and practitioners have limited time and resources, rendering early triage and prioritization of digital evidence a necessity.

As more digital data is created and digital storage systems grow in volume, digital forensic practitioners are overwhelmed by the massive amounts of data to be analyzed. Backlogs in digital forensic laboratories are common. According to a report by the FBI Regional Computer Forensics Laboratory Program [21], more than 15,000 digital devices and storage media were previewed and six petabytes of data were processed by the FBI in 2017 alone. Several Regional Computer Forensics Laboratories set the reduction of backlogs as an explicit goal.

Digital forensic practitioners seek to prioritize the data sources to be analyzed given limited resources and time. Manual and forensic-tool-based analyses may take many hours to complete for each data source. Even with automated tools such as EnCase [7], FTK and Autopsy, additional human review time is required before forensic analyses of drives can be conducted. Practitioners often do not have adequate information to make decisions about which media to work on first, something that can only be determined by spending valuable time and resources to evaluate each candidate source. The lack of efficient tools and knowledge about potential evidence in devices cause inefficiencies that can lead to critical deadlines being missed and delays in disseminating actionable information.

In the face of limited time, digital forensic practitioners must pick and choose which digital media to review from among many, rendering media triage a necessity. While triage tools exist for explicit tasks such as finding substrings of interest or specific files, a general-purpose triage method based on similarity measures between arbitrary-sized content and a labeled collection of digital media images is required. For example, a hard disk image that has high similarity to a cluster of previously-labeled drive images of interest can be prioritized for further analysis. Media similarity may also be used to infer relationships between entities, and as the basis for examining additional media and supporting the collection and analysis of additional evidence.

This chapter presents and validates a method for parallelizing the computation of a digital media similarity measure called the Jaccard index with normalized frequency (JINF) [14]. JINF, which is a variant of the Jaccard index, is based on digital media sector content comparisons. The computations adjust set intersection and union counts according to the frequencies of the set members, which are the contents of digital media sectors in the application. The JINF similarity measure applies to digital media of varying size, operating systems, filesystems and form factors.

The proposed parallelization method splits the digital media of interest into N equal partitions. Each partition is distributed to a separate processor for computation of the precursor values for the JINF similarity measure. The precursor operation is the most computationally intensive part because it involves the processing of each sector in the media to be compared. The precursor values are then passed back to a central processing node, which aggregates the results and computes the final similarity measure. The final computation is a modified version of the original JINF computation that leverages the distributed precursor values while retaining the results and accuracy of the original JINF computation.

2. Previous Work

Several digital forensic algorithms and tools use string searches as their basis. The strings may be user-specified regular expressions that match features such as email addresses, telephone numbers, social security numbers, credit card numbers, network IP addresses and other information corresponding to pseudo-unique identifiers [9, 12, 19, 29]. Garfinkel [8] defines a pseudo-unique identifier as an identifier with sufficient entropy in a given corpus that it is highly unlikely to be repeated by chance.

Garfinkel [8] also noted that hard drive images are not regularly correlated with other images. He listed three problems: (i) improper prioritization, (ii) lost opportunities for data correlation and (iii) improper emphasis on document recovery. He attempted to address these problems via cross-drive analysis using pseudo-unique information such as social security numbers, credit card numbers and email addresses. In his approach, feature extractors analyzed the extracted string files and wrote their results to feature files. The extracted features were then applied to a multi-drive corpus to identify associations between drives. The research described in this chapter also attempts to address these three problems by providing digital forensic professionals with rapid media triage and prioritization capabilities, as well as a means for identifying previously-unknown associations between digital media and the entities using the devices containing the media without reliance on document recovery.

In the case of second-order cross-drive analysis, a different question is raised: Which drives in a corpus have the largest number of features in common? To answer this question, Garfinkel [8] implemented the Multi Drive Correlator (MDC) that takes in a set of drive images with a feature to be correlated and outputs a list of (feature, drive-list) tuples. The program reads multiple feature files and generates a report that shows the number of drives in which each feature was seen, total number of times each feature was seen in the drives and list of drives in which each feature occurred.

Beverly et al. [1] extended this work using Ethernet media access control (MAC) addresses extracted from validated IP packets. They treated MAC addresses and drive images as nodes, and addresses on a hard drive image as links in a graph. They partitioned the graph to obtain distinct clusters in the collection of drive images.

Young et al. [31] introduced a file-agnostic approach that leverages hashing speed. Their approach employs sector hashes instead of file hashes. It compares blocks (fixed-sized file fragments) against a large dataset of sector hashes and considers individual sectors and collections of contiguous sectors (blocks or clusters). The approach is based on two hypotheses:

- If a block of data in a file is distinct, then a copy of the block found in a data storage device constitutes evidence that the file is or was present.

- If the blocks in a file are shown to be distinct with respect to a large and representative corpus, then the blocks can be treated as if they are universally distinct.

Young et al. [31] suggest that analyses of digital media would be more accurate and faster if a database of hash values computed from fixed-sized blocks of data is used. They employed large corpora such as Govdocs [10] and NSRL RDS [17] to populate a hash value database. Three types of sectors – singleton, paired and common sectors – were analyzed to understand the root causes of non-distinct blocks. They discovered that common sectors were typically encountered when the same blocks were present in multiple files due to malware code reuse and common file container formats.

In order to implement a field deployment on a laptop, Young and colleagues considered sampling sectors instead of processing all the media sectors. Several database implementations were considered and a Bloom filter front-end was ultimately implemented to speed up generic query times [3]. Young and colleagues analyzed several filesystems to demonstrate the generality of their approach. However, encrypted files and filesystems were found to be problematic because the (same) data of interest is stored differently when encrypted.

Moia et al. [15] assessed the impact of common blocks on similarity. They showed how common data can be identified and how the data is spread over various file types and their frequencies. They observed that common data is often generated by applications, not by users. They also demonstrated that removing common data reduced the number of matches by approximately 87%.

Garfinkel and McCarrin [11] have proposed hashing blocks instead of entire files. This block hashing method inspired the drive similarity measure methodology proposed in [14]. Garfinkel and McCarrin also specified the HASH-SETS algorithm that identifies the existence of files and the HASH-RUN algorithm that reassembles files using a database of file block hashes. A fixed block size (e.g., 4 KiB) may present a problem due to filesystem alignment. However, this is addressed by hashing overlapping blocks with a 4 KiB sliding window over the entire drive and moving the window one sector at a time.

Taguchi [27] experimented with different sample sizes using random sampling and sector hashing for drive triage. Given a drive, the goal was to provide a digital forensic practitioner with information about the utility of continuing an investigation. If a block hash value of target data is in the database, then it is very likely that the target file is on the drive. However, if no hashes are found during sampling, then a confidence level is computed to express the likelihood that the target data is not on the drive.

The spamsum program developed by Tridgell [28] performs context-triggered piecewise hashing to find updates of files. It identifies email messages that are similar to known spam. The ssdeep program [13], based on spamsum, computes and matches context-triggered piecewise hash values. It is more effective than spamsum for relatively small objects that are similar in size. However, it is vulnerable to attacks that insert trigger sequences at the beginning of files, exploiting the fact that an ssdeep signature value can have at most 64 characters [4].

Roussev et al. [23–25] have developed a similarity digest hashing method implemented in a program called sdhash. The sdhash program finds the features in a neighborhood with the lowest probability of being encountered by chance. Each selected feature, which is a 64-byte sequence, is hashed and placed in a Bloom filter. When the Bloom filter reaches full capacity, a new filter is generated. Thus, a similarity digest is a collection of a sequence of Bloom filters.

Breitinger et al. [5] developed MRSH-v2, which is based on the MRS hash [26] and context-triggered piecewise hashing. The algorithm uses a sequence of Bloom filters for fast comparison instead of a Base64-encoded fingerprint. It divides an input into chunks using a rolling hash. Each chunk is hashed and inserted into a Bloom filter. Like sdhash, MRSH-v2 has a variable length fingerprint, targeting 0.5% of the input length. Specific inputs determine whether the algorithm runs in the fragment detection and file similarity modes.

Oliver et al. [18] have proposed a locality-sensitive hashing methodology called TLSH. TLSH populates an array of bucket counts by processing an input byte sequence using a sliding window. Quartile points are computed from the array, following which digest headers and bodies are constructed. The digest header values are based on the quartile points, file length and checksum. A digest body comprises a sequence of bit pairs determined by each bucket value in relation to the quartile points. A distance score is assigned between two digests; this score is the summed-up distance between the digest headers and bodies. The distance between two digest headers is based on file lengths and quartile ratios. The distance between two digest bodies is computed as the Hamming distance. Experiments indicate that TLSH is more robust to random adversarial manipulations than ssdeep and sdhash.

Penrose et al. [20] used a Bloom filter for rapid contraband file detection. The Bloom filter reduces the size of the hash database by an order of magnitude, but incurs a small false positive rate. Penrose and colleagues subsequently implemented a larger Bloom filter for faster access, achieving 99% accuracy while scanning for contraband files in a test dataset within minutes.

Bjelland et al. [2] present three common scenarios where approximate matching can be applied: (i) search, (ii) streaming and (iii) clustering. In a search scenario, the data space is large compared with a streaming scenario. In a clustering scenario, the input and data spaces are the same. Approximate matching is impractical for large datasets due to its high latency.

Breitinger et al. [6] focus on approximate matching (i.e., similarity hashing or fuzzy hashing). They divide approximate matching methods into three main categories: (i) bytewise, (ii) syntactic and (iii) semantic matching. Bytewise matching relies only on the sequences of bytes that make up a digital object, without reference to any structures in the data stream, or to any meaning the byte stream may have when interpreted. The method described in this chapter can viewed as employing bytewise matching because it does not rely on the internal structure of a hard drive and does not give any meaning to the byte stream.

Moia and Henriques [16] have presented steps for developing new approximate matching functions. Approximate matching functions address the limitations of cryptographic hash functions that cannot detect non-identical, but similar, data.

The main goal of this research is to reduce the time required to compute JINF digital media similarity measures. Most of the methods described above can be parallelized in a manner similar to the proposed approach. As in the case of the JINF parallelization described in this chapter, most of the methods described above would also require certain modifications to be parallelized.

3. Jaccard Indexes of Similarity

This section describes the basic Jaccard index of similarity and the Jaccard index of similarity with normalized frequency.

3.1 Jaccard Index

The Jaccard index (JI) is a simple and widely-used similarity measure for arbitrary sets of data [22]. It is defined as the size of the intersection divided by the size of the union of sets A and B:

$$\mathrm{JI}(A, B) = \frac{|A \cap B|}{|A \cup B|} = \frac{|A \cap B|}{|A| + |B| - |A \cap B|} \qquad 0 \leq \mathrm{JI}(A, B) \leq 1$$

The weighted Jaccard index is used to express the similarity between two hard drives. Specifically, if $x = (x_1, x_2, \ldots, x_n)$ and $y = (y_1, y_2, \ldots, y_n)$ are two vectors with real values $x_i, y_i \geq 0$, then the weighted Jaccard index JI_w is defined as:

$$\mathrm{JI}_w = \frac{\sum_{i=1}^{n} \mathrm{Min}(x_i, y_i)}{\sum_{i=1}^{n} \mathrm{Max}(x_i, y_i)}$$

3.2 Jaccard Index with Normalized Frequency

Lim and Jones [14] introduced the basic Jaccard index with frequency (JIWF) and the Jaccard index with normalized frequency (JINF). JIWF is based on the Jaccard index and sector hashing. JINF is designed to address the sensitivity of JIWF to images of different sizes, for example, when comparing a thumb drive against a multi-terabyte hard drive.

JINF is described in detail because it is modified to parallelize the computations without affecting the numerical results. The new method, which is called the Jaccard index with split files (JISPLIT), is based on JINF with the addition of an intelligent splitting and modified computation mechanism.

In this work, the digital media of known interest is called the source drive and the media of investigatory interest is called the target drive. Parallelized computations are performed using the ARGO distributed computing platform.

The standard Jaccard index computation employs intersection and union. The computation of the new Jaccard index JINF, which is based on the weighted Jaccard index, employs modified definitions of *Intersection** (I^*) and *Union** (U^*):

$$Intersection^*(N1, N2) = \text{Min}(|N1|, |N2|)$$

$$Union^*(N1, N2) = \text{Max}(|N1|, |N2|)$$

where $N1$ and $N2$ are normalized frequencies.

The normalized frequency N_f is given by:

$$N_f = F_i/S_T$$

where F_i is the frequency of a sector hash value i and S_T is the total number of sectors in a drive.

JINF requires two normalized values to be computed for each distinct hash value, one for the source and the other for the target. As described in [14], non-probative sectors are removed before the JINF computation to improve similarity measure performance. The non-probative sectors, which are collected in a "whitelist," comprise NULL byte and SPACE byte sectors as well as sectors that appear in a clean operating system installation. The sectors written during operating system installation are not produced by user activity and, therefore, would not contribute to the similarity measure. The sector hashes computed for a clean operating system installation on test drives are saved in a database for pre-filtering (exclusion) purposes.

In general, the JINF similarity value is computed as:

$$\text{JINF}(S, T) = \frac{\text{Sum of } Intersection^*(S, T)}{\text{Sum of } Union^*(S, T)}$$

where S and T are the source and target drives, respectively.

Table 1 shows the hash values, sector frequencies and normalized sector frequencies for hypothetical source and target drives.

Table 2 shows the *Intersection** and *Union** values computed for the hypothetical source and target drives using the normalized frequency values in Table 1. The sum of *Intersection** values over all the hashes is 0.6 and the sum of all *Union** values is 1.4. The resulting JINF value is $0.6/1.4 = 0.4286$. Note that the JINF value is not dependent on which drive is the source and which drive is the target. The *Intersection** and *Union** are computed using the Min and Max of the normalized frequency of each sector hash from both drives.

Table 3 shows how the JINF values change when the frequency of sector hash A is successively increased by one in target drives $T2$, $T3$

Table 1. Hash values and frequencies of source drive S and target drive T.

Hash Value	Source Drive S Frequency	Normalized Frequency	Hash Value	Target Drive T Frequency	Normalized Frequency
A	5	0.3333	A	1	0.0667
B	4	0.2667	B	2	0.1333
C	3	0.2000	C	3	0.2000
D	2	0.1333	D	4	0.2667
E	1	0.0667	E	5	0.3333
Total	15	1	Total	15	1

Table 2. *Intersection** and *Union** of two normalized frequency values.

Hash Value	Normalized Source Frequency	Normalized Target Frequency	*Intersection** (I^*)	*Union** (U^*)
A	0.3333	0.0667	0.0667	0.3333
B	0.2667	0.1333	0.1333	0.2667
C	0.2000	0.2000	0.2000	0.2000
D	0.1333	0.2667	0.1333	0.2667
E	0.0667	0.3333	0.0667	0.3333
Total			0.6	1.4
JINF (S, T)			0.4286	

Table 3. JINF values of target drives $T2$, $T3$ and $T4$.

Hash	Freq.	$T2$ I^*	U^*	Freq.	$T3$ I^*	U^*	Freq.	$T4$ I^*	U^*
A	2	0.1250	0.3333	3	0.1764	0.3333	4	0.2222	0.3333
B	2	0.1250	0.2667	2	0.1176	0.2667	2	0.1111	0.2667
C	3	0.1875	0.2000	3	0.1764	0.2000	3	0.1667	0.2000
D	4	0.1333	0.2500	4	0.1333	0.2352	4	0.1333	0.2222
E	5	0.0667	0.3125	5	0.0666	0.2941	5	0.0667	0.2778
Sum	16	0.6375	1.3625	17	0.6705	1.3294	18	0.7	1.3
JINF	$(S, T2) = 0.4678$			$(S, T3) = 0.5044$			$(S, T4) = 0.5384$		

Table 4.   JINF values of target drives $T5$, $T6$ and $T7$.

Hash	T5			T6			T7		
	Freq.	I^*	U^*	Freq.	I^*	U^*	Freq.	I^*	U^*
A	5	0.2631	0.3333	5	0.2778	0.3333	10	0.0667	0.3333
B	2	0.1052	0.2667	2	0.1111	0.2667	20	0.1333	0.2667
C	3	0.1578	0.2000	3	0.1667	0.2000	30	0.2000	0.2000
D	4	0.1333	0.2106	4	0.1333	0.2222	40	0.1333	0.2667
E	5	0.0667	0.2631	4	0.0667	0.2222	50	0.0667	0.3333
Sum	19	0.7263	1.2736	18	0.7556	1.2444	150	0.6	1.4
JINF	$(S, T5) = 0.5702$			$(S, T6) = 0.6071$			$(S, T7) = 0.4286$		

and $T4$ (normalized frequencies of the drives are not shown). As the frequency of the first block A in the target drive moves toward the frequency of the same sector hash A in the source drive S, the similarity should increase. Each drive in Table 3 is essentially a new target drive that is checked against the source drive S. For each target drive, the JINF similarity value increases when the frequency of sector hash A increases. Note that the total number of blocks increases by one as the frequency of sector hash A is increased by one. The increase in the total number of blocks reduces the similarity because the portion of each block against the total number of blocks decreases. In contrast, the positive effect of increasing the frequency of sector hash A is greater than the negative effect of increasing the total number of blocks.

Table 4 shows how the similarity values increase when the frequency of sector hash E is incorporated in the computations. $T6$ is a new target drive created from target drive $T5$, where the frequency of sector hash E in target drive $T6$ is reduced by one (from five to four) in target drive $T5$. Target drive $T6$ has a JINF value of 0.6071, which is higher than the JINF value of 0.5702 of target drive $T5$. This is because the total number of blocks in target drive $T6$ is closer to the number in the source drive S and has less negative impact on the JINF value computation compared with target drive $T5$.

Target drive $T7$ in Table 4 demonstrates how well the method copes with a size difference between the target and source drives. The frequency of each block is copied from target drive T in the right-hand side of Table 1 and multiplied by ten. The JINF(S, T) and JINF$(S, T7)$ values are the same because the normalized frequency of each hash block is the same for both drives. Therefore, JINF does not require the sizes of the drives to be the same, or even to be measured.

Woods et al. [30] have created a realistic forensic corpus, M57-Patents, that supports cyber security and digital forensics research. The multimodal corpus includes memory and hard drive images, network packet captures and images from USB drives and cellphones.

A scripted scenario was prepared when creating the corpus. Data in the corpus is simpler than real-world data, but it is complex enough to be useful for digital forensics research. The M57-Patents corpus [10] comprises 68 hard disk images that were taken from four computer systems named Pat, Terry, Jo and Charlie over a 25-day period; each system was imaged 17 times during the 25-day experiment. Lim and Jones [14] leveraged the M57 Patents corpus to validate the JINF similarity measure. A total of 305 hard disk images (target) were compared with four disk images (source). Except for the first few images, correct results were obtained with accuracy greater than 98%. Poor results for the first few images were likely due to limited user data at the start of the M57-Patents scenario, resulting in user images with minimal differences.

4. Jaccard Index with Split Files

This section discusses the Jaccard index with split files (JISPLIT). The *Target* image is split into a number of smaller files and similarity measures are computed between the *Source* image and smaller *Target* files. Asssume that the *Target* is split into ten small files: $Trgt_{sp_1}$, $Trgt_{sp_2}$, ..., $Trgt_{sp_{10}}$.

A pool-like class in the mpi4py library (like the one in the Python multiprocessing library) is employed. The MPIPoolExecutor.map() function handles the complexity of coordinating communications with nodes, farms the tasks and collects the results. To implement parallel processing, each computation of JINF(*Source*, $Trgt_{sp_i}$) is assigned to a node in the ARGO distributed computing platform.

Table 5 shows the results of the preliminary experiments, which establish that the sum of JINF values using the split files does not match the JINF value for two complete images.

To understand the difference, consider the *Source* and *Target* images shown in Tables 6 and 7.

The *Target* image is split into two files $T1$ and $T2$ shown in Table 8. Table 9 shows the hash frequencies of the two split files.

Tables 10 and 11 show the JINF(*Source*, $T1$) and JINF(*Source*, $T2$) computations, respectively. The sum of the two JINF values is 0.22222 + 0.15 = 0.37222. However, Table 5 shows that JINF(*Source*, *Target*) has a value of 0.3043.

Table 5. JINF(*Source*, *Target*).

Normalized Hash Count	I^*	U^*
A: (1/15, 2/15)	1/15	2/15
B: (1/15, 1/15)	1/15	1/15
C: (1/15, 1/15)	1/15	1/15
D: (1/15, 1/15)	1/15	1/15
E: (1/15, 1/15)	1/15	1/15
F: (1/15, 1/15)	1/15	1/15
G: (1/15, 1/15)	1/15	1/15
H: (1/15, 0/15)	0	1/15
I: (2/15, 0/15)	0	2/15
J: (2/15, 0/15)	0	2/15
K: (3/15, 0/15)	0	3/15
Q: (0, 2/15)	0	2/15
Y: (0, 2/15)	0	2/15
Z: (0, 3/15)	0	3/15
Sum	7/15	23/15
JINF(*Source*, *Target*) $= \frac{(7/15)}{(23/15)} =$ **0.3043**		

Table 6. *Source* layout.

Sector	Hash
1	K
2	B
3	C
4	I
5	J
6	K
7	G
8	H
9	I
10	D
11	J
12	E
13	A
14	K
15	F

Table 7. *Target* layout.

Sector	Hash
1	A
2	Z
3	Q
4	A
5	B
6	C
7	D
8	E
9	F
10	G
11	Y
12	Q
13	Y
14	Z
15	Z

Table 8. Layouts of the split files $T1$ and $T2$.

T1		T2	
Sector	**Hash**	**Sector**	**Hash**
1	A	1	E
2	Z	2	F
3	Q	3	G
4	A	4	Y
5	B	5	Q
6	C	6	Y
7	D	7	Z
		8	Z

Table 9. Hash frequencies of the split files $T1$ and $T2$.

T1		T2	
Hash	**Frequency**	**Hash**	**Frequency**
A	2	E	1
B	1	F	1
C	1	G	1
D	1	Q	1
Q	1	Y	2
Z	1	Z	2

The discrepancy occurs because hash Q and hash Z are split into two different files. When a hash is split into multiple target files, the *Union** of the hash cannot have the same maximum value that it does in the non-split JINF computation. If the sum of the *Union** values is computed over all $(Source, Trgt_{sp_i})$ pairs, the *Union** value of the split hash is always less than the maximum *Union** value for $(Source, Target)$. However, if the sum of the *Intersection** values is computed over all $(Source, Trgt_{sp_i})$ pairs, the result is the same as the sum of the *Intersection** values for $(Source, Target)$.

To address this anomaly, pre-processing and post-processing must be performed before and after the ARGO platform is employed for distributed computations. The pre-processing step applies the `panda groupby` operation to the hash field, which sorts the hash values. This ensures that the same hash is placed in the same file during a split. In the example above, the hash values Q and Z had to be either in $T1$ or $T2$, both not in both files.

Table 10.   JINF(*Source, T1*).

Normalized Hash Count	I^*	U^*
A: (1/15, 2/15)	1/15	2/15
B: (1/15, 1/15)	1/15	1/15
C: (1/15, 1/15)	1/15	1/15
D: (1/15, 1/15)	1/15	1/15
E: (1/15, 0)	0	1/15
F: (1/15, 0)	0	1/15
G: (1/15, 0)	0	1/15
H: (1/15, 0)	0	1/15
I: (2/15, 0)	0	2/15
J: (2/15, 0)	0	2/15
K: (3/15, 0)	0	3/15
Q: (0, 1/15)	0	**1/15**
Y: (0, 0)	0	0
Z: (0, 1/15)	0	**1/15**
Sum	4/15	18/15

JINF(*Source, T1*) = $\frac{(4/15)}{(18/15)}$ = **0.2222**

Table 11.   JINF(*Source, T2*).

Normalized Hash Count	I^*	U^*
A: (1/15, 0)	0	1/15
B: (1/15, 0)	0	1/15
C: (1/15, 0)	0	1/15
D: (1/15, 0)	0	1/15
E: (1/15, 1/15)	1/15	1/15
F: (1/15, 1/15)	1/15	1/15
G: (1/15, 1/15)	1/15	1/15
H: (1/15, 0)	0	1/15
I: (2/15, 0)	0	2/15
J: (2/15, 0)	0	2/15
K: (3/15, 0)	0	3/15
Q: (0, 1/15)	0	**1/15**
Y: (0, 2/15)	0	2/15
Z: (0, 2/15)	0	**2/15**
Sum	3/15	20/15

JINF(*Source, T2*) = $\frac{(3/15)}{(20/15)}$ = **0.15**

After the ARGO platform completes its parallel processing for each (*Source, Trgt$_{sp_i}$*) pair, each ARGO node returns the following two-tuple:

$$(value_i, \{(h1, NF(h1)), (h2, NF(h2)), \ldots, (hk, NF(hk))\}_i)$$

where $NF(h)$ is the normalized frequency of a hash h and k is the total number of unique hashes assigned to each ARGO node i.

The first element of a tuple is the sum of the *Intersection** values (Table 10 or 11) and the second is the *Union** column with the corresponding hash values (Table 10 or 11). A separate ARGO node is assigned to perform the computations of JINF(*Source, Trgt$_{sp_i}$*) for each pair. Note that an ARGO node does not complete the normal JINF computations. Instead, it returns an intermediate result as a two-tuple:

$$\text{JISPLIT}(Source, Target) = \frac{\text{Final sum of } Intersection^*}{\text{Final sum of } Union^*}$$

where

$$\text{Final sum of } Intersection^* = \sum_{i=1}^{N} \text{Sum of } Intersection^* \text{ from } node_i$$

and

Table 12. Hash frequencies of the split files from *Target*.

T_a		T_b	
Hash	**Frequency**	**Hash**	**Frequency**
A	2	G	1
B	1	Q	2
C	1	Y	2
D	1	Z	3
E	1		
F	1		

$$\text{Final sum of } Union^* = \sum_{j=1}^{H} Union^*_j \text{ row in final } Union^* \text{ column}$$

where N is the number of split files and H is the number of unique hashes.

The sum of *Intersection** values is the numerator in the final JIS-PLIT(*Source, Target*) computation, which is straightforward. In the case of the denominator in the final JINF computation, the central node is used to construct the final *Union** column using the *Union** columns returned the ARGO nodes.

The normalized frequency of each hash value h in the final *Union** column is Max($NF(h)_1$, $NF(h)_2$, ..., $NF(h)_H$), where $NF(h)_i$ is the normalized frequency of hash h from ARGO node i. The sum of all the rows in the final *Union** column is the denominator in the final JISPLIT(*Source, Target*) computation.

An example is helpful to illustrate the computations. The *Target* image is split into two files, T_a and T_b, by applying the **groupby** operation on the hash field of the *Target*. Table 12 shows the hash frequencies of the split files from *Target*.

Assume that ARGO $node_a$ computes JINF(*Source*, T_a) and ARGO $node_b$ computes JINF(*Source*, T_b). The $node_a$ returns 6/15 as the sum of *Intersection** along with the *Union** column, which is the last column in Table 13. The $node_b$ returns 1/15 as the sum of *Intersection** along with the *Union** column, which is the last column in Table 14.

The two-tuple returned by $node_a$ is (6/15, {(A, 2/15), (B, 1/15), (C, 1/15), ..., (K, 3/15)}). As shown in Table 15, the final sum of *Intersection** obtained via post-processing is computed as the sum of (6/15, 1/15), which is 7/15.

Table 13.   JINF(*Source*, T_a).

Normalized Hash Count	I^*	U^*
A: (1/15, 2/15)	1/15	2/15
B: (1/15, 1/15)	1/15	1/15
C: (1/15, 1/15)	1/15	1/15
D: (1/15, 1/15)	1/15	1/15
E: (1/15, 1/15)	1/15	1/15
F: (1/15, 1/15)	1/15	1/15
G: (1/15, 0)	0	1/15
H: (1/15, 0)	0	1/15
I: (2/15, 0)	0	2/15
J: (2/15, 0)	0	2/15
K: (3/15, 0)	0	3/15
Q: (0, 0)	0	0
Y: (0, 0)	0	0
Z: (0, 0)	0	0
Sum	6/15	

Table 14.   JINF(*Source*, T_b).

Normalized Hash Count	I^*	U^*
A: (1/15, 0)	0	1/15
B: (1/15, 0)	0	1/15
C: (1/15, 0)	0	1/15
D: (1/15, 0)	0	1/15
E: (1/15, 0)	0	1/15
F: (1/15, 0)	0	1/15
G: (1/15, 1/15)	1/15	1/15
H: (1/15, 0)	0	1/15
I: (2/15, 0)	0	2/15
J: (2/15, 0)	0	2/15
K: (3/15, 0)	0	3/15
Q: (0, 2/15)	0	2/15
Y: (0, 2/15)	0	2/15
Z: (0, 3/15)	0	3/15
Sum	1/15	

Table 15.   Final *Intersection** from two ARGO nodes.

Pair of Split Files	I^*
(*Source*, T_a)	6/15
(*Source*, T_b)	1/15
Final Sum	7/15

As shown in Table 16, the final *Union** column is constructed by selecting the maximum of the two columns returned by the two ARGO nodes. The denominator value is the sum of the final *Union** column, which is 23/15. The resulting JISPLIT value is $\frac{7/15}{23/15} = 0.3043$, which matches the JINF(*Source*, *Target*) value in Table 5.

At this point, only the *Target* image is split. The *Source* image may also be split as shown in Table 17. The same pre-processing step is applied.

Four ARGO nodes are required because there are four pairs of split file combinations: (S_a, T_a), (S_a, T_b), (S_b, T_a) and (S_b, T_b). Tables 18 through 21 show the JINF computations for the four pairs.

Table 16. Final *Union** column.

Hash	U^* (*Source, T_a*)	U^* (*Source, T_b*)	Final U^* Column
A	**2/15**	1/15	Max(2/15, 1/15) = **2/15**
B	1/15	1/15	1/15
C	1/15	1/15	1/15
D	1/15	1/15	1/15
E	1/15	1/15	1/15
F	1/15	1/15	1/15
G	1/15	1/15	1/15
H	1/15	1/15	1/15
I	2/15	2/15	2/15
J	2/15	2/15	2/15
K	3/15	3/15	3/15
Q	0	2/15	2/15
Y	0	2/15	2/15
Z	0	3/15	3/15
Final Sum	1/15	22/15	**23/15**

Table 17. Hash frequencies of the split files of *Source*.

S_a		S_b	
Hash	Frequency	Hash	Frequency
A	1	F	1
B	1	G	1
C	1	H	1
D	1	I	2
E	1	J	2
		K	3

Table 22 shows the *Intersection** values returned by the four ARGO nodes. The final sum of 7/15 is the numerator in the JISPLIT(*Source, Target*) computation.

Table 23 shows the final *Union** columns returned by the four ARGO nodes. The final sum of 23/15 is the denominator in the JISPLIT(*Source, Target*) computation. The JISPLIT value obtained from the four ARGO nodes is $\frac{7/15}{23/15} = 0.3043$, which matches the JINF(*Source, Target*) value in Table 5.

Table 19. JINF(S_b, T_a).

Normalized Hash Frequency	I^*	U^*
A: (0, 2/15)	0	2/15
B: (0, 1/15)	0	1/15
C: (0, 1/15)	0	1/15
D: (0, 1/15)	0	1/15
E: (0, 1/15)	0	1/15
F: (1/15, 1/15)	1/15	1/15
G: (1/15, 0)	0	1/15
H: (1/15, 0)	0	1/15
I: (2/15, 0)	0	2/15
J: (2/15, 0)	0	2/15
K: (3/15, 0)	0	3/15
Sum	1/15	

Table 18. JINF(S_a, T_a).

Normalized Hash Frequency	I^*	U^*
A: (1/15, 2/15)	1/15	2/15
B: (1/15, 1/15)	1/15	1/15
C: (1/15, 1/15)	1/15	1/15
D: (1/15, 1/15)	1/15	1/15
E: (1/15, 1/15)	1/15	1/15
F: (0, 1/15)	0	1/15
Sum	5/15	

Table 20. JINF(S_a, T_b).

Normalized Hash Frequency	I^*	U^*
A: (1/15, 0)	0	1/15
B: (1/15, 0)	0	1/15
C: (1/15, 0)	0	1/15
D: (1/15, 0)	0	1/15
E: (1/15, 0)	0	1/15
G: (0, 1/15)	0	1/15
Q: (0, 2/15)	0	2/15
Y: (0, 2/15)	0	2/15
Z: (0, 3/15)	0	3/15
Sum	0	

Table 21. JINF(S_b, T_b).

Normalized Hash Frequency	I^*	U^*
F: (1/15, 0)	0	1/15
G: (1/15, 1/15)	1/15	1/15
H: (1/15, 0)	0	1/15
I: (2/15, 0)	0	2/15
J: (2/15, 0)	0	2/15
K: (3/15, 0)	0	3/15
Q: (0, 2/15)	0	2/15
Y: (0, 2/15)	0	2/15
Z: (0, 3/15)	0	3/15
Sum	1/15	

5.   Results and Validation

The computing performance of JISPLIT using the ARGO distributed environment was evaluated. The M57 Patents Scenario dataset was used for initial performance testing. The daily images for Terry were 40 GB each and the system images for Pat, Jo and Charlie were 10 GB each. For the 10 GB source image files and 10 GB target image files, the processing time improved around 15%, from an average of 13 minutes

Table 22. Final *Intersection** from four ARGO nodes.

Pair of Split Files	I^*
(S_a, T_a)	5/15
(S_b, T_a)	1/15
(S_a, T_b)	0/15
(S_b, T_b)	1/15
Final Sum	**7/15**

Table 23. Final *Union** column from four ARGO nodes.

Hash	U^* (S_a, T_a)	U^* (S_b, T_a)	U^* (S_a, T_b)	U^* (S_b, T_b)	Final U^* Column
A	2/15	2/15	1/15	N/A	2/15
B	1/15	1/15	1/15	N/A	1/15
C	1/15	1/15	1/15	N/A	1/15
D	1/15	1/15	1/15	N/A	1/15
E	1/15	1/15	1/15	N/A	1/15
F	1/15	1/15	1/15	1/15	1/15
G	N/A	1/15	1/15	1/15	1/15
H	N/A	1/15	1/15	1/15	1/15
I	N/A	2/15	2/15	2/15	2/15
J	N/A	2/15	2/15	2/15	2/15
K	N/A	3/15	3/15	3/15	3/15
Q	N/A	N/A	2/15	2/15	2/15
Y	N/A	N/A	2/15	2/15	2/15
Z	N/A	N/A	3/15	3/15	3/15
Final Sum					**23/15**

to an average of 11 minutes. When the number of split image files was increased to more than a certain value that depended on the original size of the pre-split image file, more time was required for the JINF computations compared with the time required without splitting. The penalty is imposed by the pre- and post-processing steps, which involve multiple file access operations.

For the 10 GB source image files and 40 GB target image files, the JIS-PLIT computation time was reduced by about half (51%) as shown in Table 24. Computing the JINF values for two huge images is computationally intensive, especially with respect to system memory. However, the results suggest that partitioning the source and target media into smaller files and distributing the computations render the overall times

Table 24. JISPLIT performance using the ARGO platform.

Target Split Files	Processing Time
1	22 min 15 sec
3	12 min 41 sec
6	**10 min 49 sec**
9	11 min 57 sec
12	12 min 29 sec

Source: 10 GB, *Target*: 40 GB

for computing similarity measures feasible. Of course, additional experimentation and testing are necessary to further validate the method.

6.   Conclusions

Early triage and prioritization of digital evidence are important tasks performed by digital forensic practitioners. The main goal of this research has been to compute digital media image similarity measures that support efficient triage and association discovery. The proposed method does not replace existing approximate hashing and other techniques; instead, it leverages and potentially augments them. Most existing similarity measures work at the file or object levels. In contrast, the proposed method works at the sector level and is, therefore, robust in the face of deleted and partially-overwritten data.

By splitting digital media images into several files via centralized pre-processing, distributing the separate file computations to nodes and computing a final similarity measure via centralized post-processing, the overall computation time can be reduced up to 51% while achieving the accuracy of the computations performed by a single processor. When the target and source images are much larger, using optimal numbers of splits for the images can yield computational time reductions exceeding 51%. The determination of the optimal number of splits for a digital media image of a given size is a topic for future research; in fact, the optimization may also depend on media content. In any case, high-performance parallel computing infrastructures may be used to implement the proposed method to maximize computational time reductions.

However, there are some caveats. The proposed method may be vulnerable to an adversary who selectively deletes or overwrites content in common with another digital device, plants false fragments to mislead the algorithm and digital forensic practitioners, wipes digital media at a low level and/or encrypts media with unique keys. While these actions would severely limit the effectiveness of the proposed method, the first two require considerable investments of time and skill on the part of an adversary and may also be detectable after the fact. The last two actions are effective, but they would also be obvious to a digital forensic practitioner and would, therefore, be more likely to lead to no results instead of false results. It should be noted that the proposed method is applicable to compressed and encrypted files as long as the compression methods and encryption keys are the same across systems. Additionally, the method is applicable to damaged media and partially-recovered content because it does not require filesystem information.

References

[1] R. Beverly, S. Garfinkel and G. Cardwell, Forensic carving of network packets and associated data structures, *Digital Investigation*, vol. 8(S), pp. S78–S89, 2011.

[2] P. Bjelland, K. Franke and A. Arnes, Practical use of approximate hash-based matching in digital investigations, *Digital Investigation*, vol. 11(S1), pp. S18–S26, 2014.

[3] B. Bloom, Space/time trade-offs in hash coding with allowable errors, *Communications of the ACM*, vol. 13, pp. 422–426, 1970.

[4] F. Breitinger and H. Baier, Performance issues about context-triggered piecewise hashing, *Proceedings of the International Conference on Digital Forensics and Cyber Crime*, pp. 141–155, 2012.

[5] F. Breitinger and H. Baier, Similarity-preserving hashing: Eligible properties and a new algorithm MRSH-v2, *Proceedings of the Fourth International Conference on Digital Forensics and Cyber Crime*, pp. 167–182, 2012.

[6] F. Breitinger, B. Guttman, M. McCarrin, V. Roussev and D. White, Approximate Matching: Definition and Terminology, NIST Special Publication 800-168, National Institute of Standards and Technologies, Gaithersburg, Maryland, 2014.

[7] S. Bunting and W. Wei, *EnCase Computer Forensics: The Official EnCE: EnCase Certified Examiner Study Guide*, Wiley Publishing, Indianapolis, Indiana, 2006.

[8] S. Garfinkel, Forensic feature extraction and cross-drive analysis, *Digital Investigation*, vol. 3(S), pp. S71–S81, 2006.

[9] S. Garfinkel, Digital media triage with bulk data analysis and bulk_extractor, *Computers and Security*, vol. 32, pp. 56–72, 2013.

[10] S. Garfinkel, P. Farrell, V. Roussev and G. Dinolt, Bringing science to digital forensics with standardized forensic corpora, *Digital Investigation*, vol. 6(S), pp. S2–S11, 2009.

[11] S. Garfinkel and M. McCarrin, Hash-based carving: Searching media for complete files and file fragments with sector hashing and `hashdb`, *Digital Investigation*, vol. 14(S1), pp. S95–S105, 2015.

[12] S. Garfinkel, A. Nelson, D. White and V. Roussev, Using purpose-built functions and block hashes to enable small block and sub-file forensics, *Digital Investigation*, vol. 7(S), pp. S13–S23, 2010.

[13] J. Kornblum, Identifying almost identical files using context-triggered piecewise hashing, *Digital Investigation*, vol. 3(S), pp. 91–97, 2006.

[14] M. Lim and J. Jones, A digital media similarity measure for triage of digital forensic evidence, in *Advances in Digital Forensics XVI*, G. Peterson and S. Shenoi (Eds.), Springer, Cham, Switzerland, pp. 111–135, 2020.

[15] V. Moia, F. Breitinger and M. Henriques, The impact of excluding common blocks in approximate matching, *Computers and Security*, vol. 89, article no. 101676, 2020.

[16] V. Moia and M. Henriques, A comparative analysis of similarity search strategies for digital forensic investigations, *Proceedings of the Thirty-Fifth Brazilian Symposium on Telecommunications and Signal Processing*, pp. 462–466, 2017.

[17] National Institute of Standards and Technology, National Software Reference Library (NSRL), Gaithersburg, Maryland (`www.nsrl.nist.gov`), 2019.

[18] J. Oliver, C. Cheng and Y. Chen, TLSH – A locality sensitive hash, *Proceedings of the Fourth Cybercrime and Trustworthy Computing Workshop*, pp. 7–13, 2013.

[19] H. Parsonage, Computer Forensics Case Assessment and Triage – Some Ideas for Discussion (`computerforensics.parsonage.co.uk/triage/ComputerForensicsCaseAssessmentANDTriageDiscussionPaper.pdf`), 2009.

[20] P. Penrose, W. Buchanan and R. Macfarlane, Fast contraband detection in large capacity disk drives, *Digital Investigation*, vol. 12(S1), pp. S22–S29, 2015.

[21] RCFL National Program Office, Regional Computer Forensics Laboratory Annual Report for Fiscal Year 2017, Quantico, Virginia (`www.rcfl.gov/file-repository/09-rcfl-annual-2017-190130-print-1.pdf/view`), 2017.

[22] R. Real and J. Vargas, The probability basis of Jaccard's index of similarity, *Systematic Biology*, vol. 45(30), pp. 380–385, 1996.

[23] V. Roussev, Building a better similarity trap with statistically improbable features, *Proceedings of the Forty-Second Hawaii International Conference on System Sciences*, 2009.

[24] V. Roussev, Data fingerprinting with similarity digests, in *Advances in Digital Forensics VI*, K. Chow and S. Shenoi (Eds.), Springer, Berlin Heidelberg, Germany, pp. 207–226, 2010.

[25] V. Roussev, Y. Chen, T. Bourg and G. Richard, `md5bloom`: Forensic filesystem hashing revisited, *Digital Investigation*, vol. 3(S), pp. S82–S90, 2006.

[26] V. Roussev, G. Richard and L. Marziale, Multi-resolution similarity hashing, *Digital Investigation*, vol. 4(S), pp. S105–S113, 2007

[27] J. Taguchi, Optimal Sector Sampling for Drive Triage, M.S. Thesis, Department of Computer Science, Naval Postgraduate School, Monterey, California, 2013.

[28] A. Tridgell, `spamsum` (`samba.org/ftp/unpacked/junkcode/spamsum/README`), 2002.

[29] R. Walls, E. Learned-Miller and B. Levine, Forensic triage for mobile phones with DEC0DE, *Proceedings of the Twentieth USENIX Conference on Security*, 2011.

[30] K. Woods, C. Lee, S. Garfinkel, D. Dittrich, A. Russell and K. Kearton, Creating realistic corpora for security and forensic education, *Proceedings of the ADFSL Conference on Digital Forensics, Security and Law*, pp. 123–134, 2011.

[31] J. Young, K. Foster, S. Garfinkel and K. Fairbanks, Distinct sector hashes for target file detection, *IEEE Computer*, vol. 45(12), pp. 28–35, 2012.

Chapter 5

EVALUATION OF NETWORK TRAFFIC ANALYSIS USING APPROXIMATE MATCHING ALGORITHMS

Thomas Göbel, Frieder Uhlig and Harald Baier

Abstract   Approximate matching has become indispensable in digital forensics as practitioners often have to search for relevant files in massive digital corpora. The research community has developed a variety of approximate matching algorithms. However, not only data at rest, but also data in motion can benefit from approximate matching. Examining network traffic flows in modern networks, firewalls and data loss prevention systems are key to preventing security compromises.

This chapter discusses the current state of research, use cases, validations and optimizations related to applications of approximate matching algorithms to network traffic analysis. For the first time, the efficacy of prominent approximate matching algorithms at detecting files in network packet payloads is evaluated, and the best candidates, namely TLSH, ssdeep, mrsh-net and mrsh-cf, are adapted to this task. The individual algorithms are compared, strengths and weaknesses highlighted, and detection rates evaluated in gigabit-range, real-world scenarios. The results are very promising, including a detection rate of 97% while maintaining a throughput of 4 Gbps when processing a large forensic file corpus. An additional contribution is the public sharing of optimized prototypes of the most promising algorithms.

Keywords: Network traffic analysis, approximate matching, similarity hashing

1.   Introduction

Data loss prevention systems protect enterprises from intellectual property and sensitive data theft. They have become almost indispensable as legal requirements like the General Data Protection Regulation (GDPR) in the European Union levy high fines when consumer data is accessed by unauthorized parties.

© IFIP International Federation for Information Processing 2021
Published by Springer Nature Switzerland AG 2021
G. Peterson and S. Shenoi (Eds.): Advances in Digital Forensics XVII, IFIP AICT 612, pp. 89–108, 2021.
https://doi.org/10.1007/978-3-030-88381-2_5

In October 2020, the German software company Software AG was hit with a ransom demand of 23 million euros in exchange for the decryption password and the promise not to disclose 1 TB of internal company documents and customer data [19]. Software AG products are used by 70% of Fortune 1,000 companies. In addition to the ransom, recovery costs and loss of reputation incurred by Software AG, the damage extended to the enterprise environments of hundreds of its customers.

In June 2016, Russian hackers using the Guccifer 2.0 pseudonym stole tens of thousands of email and documents, including 8,000 attachments from the U.S. Democratic National Committee. The public release of the stolen documents on WikiLeaks constituted an attempt to actively influence the 2016 U.S. presidential election [16].

The two incidents demonstrate the costs of data protection failures. Also, they underscore the importance of checking data flows at transition points in networks. Whether this is accomplished using intrusion detection/prevention systems or data loss prevention systems, transparency at network transition points is vital to protecting sensitive data. Enterprise networks host a variety of platforms for data and file exchange. Emerging technologies, such as cloud platforms and workstream collaboration platforms (e.g., Slack and Mattermost), simplify data exchange in enterprises. However, they increase the risk of sensitive data leaving the premises of controlled networks and falling into the wrong hands.

To reduce the risk of data leakage, data loss prevention solutions must be implemented in networks, at network endpoints and in the cloud. A single data loss protection solution is ineffective; rather, a multi-tiered approach is needed to protect an enterprise infrastructure from unauthorized data access. Indeed, an orchestrated solution involving endpoints, storage and networks is advised [1]. Such a solution filters data at rest, data in use and data in motion to prevent the data from leaving the virtual premises of an enterprise. Since most data loss prevention systems are closed-source solutions, without internal knowledge of these products, it is difficult to judge their protection scopes and effectiveness.

Most commercial data loss protection solutions fall into one or combinations of three categories:

- **Endpoint Protection:** These solutions involve desktop and/or server agents that enable security teams to monitor data at endpoints. Predefined rules and blacklists prevent users from copying data (in use) to removable devices or transferring data to unauthorized web destinations.

- **Storage Protection:** These solutions focus on securing data at rest. Several cloud-based solutions fall in this category. Many of them protect intellectual property on Amazon AWS or Google Cloud platforms. They achieve their goals in part through access restrictions and blacklisting.

- **Network Protection:** These solutions protect data in motion, specifically at the ingress and egress points of enterprise networks. Many solutions prevent leakage by analyzing network packet metadata. Although network packet content analysis is hindered by encryption (approximately 80% of traffic is currently encrypted and the trend is towards total encryption), it is feasible for enterprises to place these solutions in-line with their network appliances. Products such as Sophos XG Firewall, Symantec Data Loss Prevention and Fortinet's Fortigate DLP integrate decryption functionality of TLS 1.3 communications. These are the network points at which the research discussed in this chapter is directly applicable. Decryption is necessary because encrypted packets are uniformly distributed and lack the recognizable features needed to leverage approximate matching.

This research focuses on preventing data loss in network environments. Approximate matching can be used to identify files in unencrypted network packet payloads. The goal is to transform theoretical concepts into reality and craft feasible solutions for identifying files with high accuracy while maintaining low failure rates at gigabit throughputs. Note that the approach does not involve prior inspection of unencrypted payloads to determine relationships between packets via string matching. Deep packet inspection methods can detect the transmission of sensitive content hidden in files. Since the majority of data loss occurs not via complete files but through portions of files being reformatted within another file or context (e.g., the Open Document Format compresses the actual contents), deep packet inspection aids data loss prevention via pattern matching and may find compressed portions of mixed files. However, these measures are not employed in this work.

This research demonstrates how file recognition in network traffic can be improved via approximate matching. The performance of prominent approximate matching algorithms is evaluated. First, file detection rates in an idealized setting are assessed without added noise from live traffic. The detection rates and maximum throughputs of the algorithms are measured. The best-performing algorithm is subsequently applied to live traffic and its performance is examined in detail. All the algorithms were adapted for packet filtering in that they were connected to a network interface and invoked whenever a packet was picked up by the interface.

Reasons for the variations in the true positive and false positive detection rates between the approximate matching algorithms are discussed. In fact, this work is the first to use prominent approximate matching algorithms to match files in real-world network traffic. Additionally, the two best approximate matching algorithms are adapted for efficient and effective real-time filtering of network traffic.

2.   Foundations and Related Work

Approximate matching was first employed in digital forensics in the mid-2000s. The U.S. National Institute of Standards and Technology (NIST) [8] defines approximate matching as "a promising technology designed to identify similarities between two digital artifacts ... to find objects that resemble each other or to find objects that are contained in another object." Approximate matching algorithms achieve this goal using three approaches [23]:

- **Bytewise Matching:** This type of matching, referred to as fuzzy hashing or similarity hashing, operates at the byte level and only takes byte sequences as inputs.

- **Syntactic Matching:** This type of matching takes bytes as inputs but also relies on internal structure information of the subjects that are intended to be matched (e.g., header information in packets may be ignored).

- **Semantic Matching:** This type of matching, which focuses on content-visual differences, resembles human recognition. For example, JPG and PNG images may have similar content (i.e., pictures), but their filetypes and byte streams are different.

Four types of approximate matching approaches are relevant to this work [25]:

- **Context-Triggered Piecewise Hashing (CTPH):** This approach locates content markers (contexts) in binary data. It computes the hash of each document fragment delimited by contexts and stores the resulting sequence of hashes. In 2006, Kornblum [24] created `ssdeep`, one of the first algorithms that computed context-triggered piecewise signatures. The algorithm produces a match score from 0 to 100 that is interpreted as a weighted measure of file similarity, where a higher score implies greater similarity.

- **Block-Based Hashing (BBH):** This approach generates and stores cryptographic hashes of fixed-size blocks (e.g., 512 bytes).

The block-level hashes of two inputs are compared by counting the number of common blocks and computing a measure of similarity. An example implementation is dcfldd [22], which divides input data into blocks and computes their cryptographic hash values. The approach is computationally efficient but it is highly unstable because adding or deleting even a single byte at the beginning of a file changes all the block hashes.

- **Statistically-Improbable Features (SIF):** This approach identifies a set of features in each examined object and compares the features; a feature in this context is a sequence of consecutive bytes selected according to some criteria from the file in which the object is stored. Roussev [30] used entropy to find statistically-improbable features. His sdhash algorithm [31] generates a score between 0 and 100 to express the confidence that two data objects have commonalities.

- **Block-Based Rebuilding (BBR):** This approach uses external auxiliary data corresponding to randomly-, uniformly- or fixed-selected blocks (e.g., binary blocks) of a file, to reconstruct the file. It compares the bytes in the original file with those in the selected blocks and computes the differences between them using the Hamming distance or another metric. The differences are used to find similar data objects. Two well-known block-based rebuilding algorithms are bbHash [5] and SimHash [34].

- **Locality-Sensitive Hashing (LSH):** This approach from the data mining field is technically not an approximate matching approach. It is considered because it is often employed in a similar context as approximate matching.

Locality-sensitive hashing is a general mechanism for nearest neighbor search and data clustering whose performance strongly relies on the hashing method. An example is the TLSH algorithm [26], which processes an input byte sequence using a sliding window to populate an array of bucket counts and determines the quartile points of the bucket counts. A fixed-length digest is constructed, which comprises a header with the quartile points, input length and checksum, and a body comprising a sequence of bit pairs that depend on each bucket's value in relation to the quartile points. The distance between two digest headers is determined by the differences in file lengths and quartile ratios. The bodies are compared using their approximate Hamming distance. The similarity score is computed based on the distances between the two headers and the two bodies.

2.1   Current State of Approximate Matching

In 2014, Breitinger and Baggili [4] proposed an approximate matching algorithm for filtering relevant files in network packets. The algorithm divides input files into 1,460 byte sequences, which simulates packets with maximum transmission units of 1,500 bytes less 20 bytes each for the IP and TCP headers. It achieved false positive rates between 10^{-4} and 10^{-5} for throughputs exceeding 650 Mbps.

The algorithm of Breitinger and Baggili was evaluated using simulated network traffic [4]. In contrast, this research has tested algorithms using real network traffic to understand their real-world applicability; additionally, this work has conducted experiments with variable packet sizes. Another key issue is that the throughput of 650 Mbps reported by Breitinger and Baggili was achieved without parallelization. In fact, they noted that parallelization could increase the speed significantly because hashing packets and performing comparisons with a Bloom filter can be done in an unsynchronized manner [10].

Since 2014, there has been little progress in advancing approximate matching in the network context. This is largely due to the encryption of network traffic payloads, which prevents file detection.

However, significant improvements have been made to approximate matching algorithms. Cuckoo filters, which are more practical than Bloom filters [17], have been adapted to approximate matching in digital forensic applications [21]. Researchers have also proposed improvements to Bloom and cuckoo filters such as XOR filters that are faster and smaller than Bloom filters [20], Morton filters that are faster, compressed cuckoo filters [11] and hierarchical Bloom filter trees that are improved Bloom filter trees [28]. However, it remains to be seen if the application of these improved filters to approximate matching will provide better file recognition performance.

2.2   Approximate Matching Algorithms

Filters are constantly being improved by the research community. However, approximate matching algorithms that employ filters and underlying matching methods advance more slowly. The following approximate matching algorithms are used frequently:

- **MRSH:** Algorithms in the multi-resolution similarity hash (MRSH) family are based on `ssdeep` and, thus, employ context-triggered piecewise hashing. Roussev et al. [33] created the original `mrshash` algorithm, which was subsequently improved to MRSH-v2, a faster version [6]. The algorithm also provides a fragment detection mode and the ability to compute file similarity.

Most relevant to this research are two variants of MRSH-v2. The mrsh-net algorithm [4] is a special version of MRSH-v2 created for network packet filtering; it is also the first approximate matching algorithm to be evaluated in a network context. The mrsh-cf algorithm [21] is a special version of mrsh-net that replaces the Bloom filter with a cuckoo filter, providing runtime improvements and much better false positive rates.

The latest variant of the MRSH family, mrsh-hbft, uses hierarchical Bloom filter trees instead of conventional Bloom filters. Lillis et al. [28] have demonstrated that the hierarchical Bloom filter trees used in mrsh-hbft improve on the original MRSH-v2 algorithm, which uses a standard Bloom filter. However, as discussed later, the mrsh-cf algorithm is still unmatched in terms of speed and accuracy, which is especially important in network applications.

- **mvHash-B:** The mvHash-B algorithm [3] employs block-based and similarity-preserving hashing, but also relies on majority voting and Bloom filters. It is highly efficient with very low runtime complexity and small digest size. The algorithm exhibits weaknesses to active adversaries [13], but these issues are now addressed [13]. A caveat is that the algorithm must be adjusted for every filetype. The original algorithm only applied to JPG and DOC files.

- **FbHash:** The FbHash algorithm [12] builds on the mvHash-B algorithm. However, its source code is not published, so the algorithm cannot be adapted to network traffic analysis. Moreover, the algorithm is limited to a small set of filetypes.

- **sdhash:** Roussev [31] developed the sdhash algorithm four years after the release of ssdeep. The sdhash algorithm uses hashed similarity digests as a means of comparison. A similarity digest contains statistically-improbable features in a file. The sdhash algorithm has been thoroughly evaluated against its predecessor ssdeep [7]. It detects correlations with finer granularity than ssdeep, which makes it a viable candidate [27]. However, a runtime comparison by Breitinger et al. [9] has demonstrated that MRSH-v2 outperforms sdhash by a factor of eight. MRSH-v2 is chosen over sdhash in this research because of its superior runtime performance and similar ability to correlate small fragments.

- **TLSH:** TLSH developed by Oliver et al. [26] is an adaptation of the Nilsimsa hash [15] used for spam detection. As mentioned above, TLSH is not an approximate matching algorithm because it is based on locality-sensitive hashing that does not conform to the approximate matching definition [8]. TLSH uses the Hamming distance between Bloom filters to compare hashes and provides similarity scores ranging from 0 to 1,000. TLSH has worse performance than ssdeep and sdhash, with sdhash consistently recognizing files at the smallest granularity. TLSH is extremely robust to random manipulations such as insertions and deletions in a file. However, it produces high false positive rates and is slightly slower than mrsh-net.

- **ssdeep:** The ssdeep algorithm is commonly used in digital forensics. It is implemented on several platforms [35] and is the *de facto* standard in some cyber security areas [29]. Applications such as VirusTotal are based on ssdeep. The algorithm generates a signature file that depends on the actual file content. It is used to compare two signature files or a signature file against a data file; the results are the same in both cases and the use of one or other method depends on the available data. Although ssdeep is a key achievement in similarity detection and is still relatively up-to-date, some limitations have been identified recently, for which certain enhancements and alternative theoretical approaches have been suggested [2].

Table 1 presents a detailed evaluation of approximate matching algorithms for network traffic analysis. Note that the mrsh-cf algorithm is the 2020 version with a newer cuckoo filter.

3.   Controlled Study

As discussed in Section 2.2, based on their reliability and performance, three algorithms are best suited to network traffic analysis: (i) ssdeep, (ii) TLSH and (iii) MRSH (mrsh-net and mrsh-cf variants). The performance, speed and applicability of the algorithms were first evaluated with respect to data at rest. Next, the three algorithms were evaluated on their ability to deal with live network packets. In the case of the MRSH family, the mrsh-net variant was evaluated as the first algorithm capable of filtering network traffic whereas the mrsh-cf was evaluated as a faster and more reliable version of mrsh-net.

The algorithms were evaluated using the well-known *t5-corpus*, which contains 4,457 files with a total size of 1.8 GB [32]. The average file size is almost 420 KiB. Table 2 shows the composition of the *t5-corpus*. Note

Table 1. Evaluations of approximate matching algorithms.

	mrsh-cf	mrsh-net	mvHash-B	sdhash	TLSH	ssdeep	FbHash	mrsh-hbft
Speed	37% runtime improvement over mrsh-net	Faster than mrsh-v2	Lowest runtime complexity	No current benchmarks	No current benchmarks	Twice as fast as TLSH (2017 version)	No current benchmarks	No current benchmarks
Fragment Detection	Small	Small	NA	5%	Small	20 to 50% of original	1%	NA
Remarks	Fastest MRSH algorithm	Special version of mrsh-v2 for network traffic	Limited to a few file types	MRSH algorithms are more effective, but sdhash is slightly more accurate	Used for file similarity instead of file identification	Cannot hash files over 2 GB	Unpublished code and limited to DOC file types	Preferable in cases with memory limitations
Basis	Context-triggered piecewise hashing, Cuckoo filter	Context-triggered piecewise hashing, Bloom filter	Block-based hashing	Statistically-improbable features	Locality-sensitive hashing	Context-triggered piecewise hashing	Term frequency inverse document frequency, similar to mvHash-B	Context-triggered piecewise hashing, Hierarchical Bloom filter
Latest Version	2020	2015	2012	2013	2020	2017	2018	2018

Table 2.   *t5-corpus* composition.

JPG	GIF	DOC	XLS	PPT	PDF	TXT	HTML
362	67	533	250	368	1,073	711	1,093

that the *t5-corpus* is a subset of the *GovDocs* corpus, which was created by crawling U.S. Government websites [18]. Due to the data collection process, it is assumed that the corpus has multiple related files.

Table 3.   Time requirements for filter generation and application.

	mrsh-cf	mrsh-net	ssdeep	TLSH	mrsh-hbft
Filter Generation	12.51 s	32.90 s	14.90 s	17.18 s	274 s
All vs. All	12.94 s	67.84 s	27.37 s	78.29 s	300 s

3.1   All vs. All Evaluation

The All vs. All evaluation sets algorithm performance baselines using data at rest. Each algorithm first generated a filter of the *t5-corpus*. Next, each algorithm with its filter was given the entire *t5-corpus* (1.8 GB) to process. Table 3 shows the time requirements for the algorithms to generate the filter and apply it to every file in the corpus.

Speed is a key indicator of the suitability of an algorithm for detecting files in network traffic. This is why the performance of each algorithm was evaluated when matching the *t5-corpus* with itself. Three of the algorithms are capable of performing small fragment detection, which is important for file matching in network packets. ssdeep is limited in this regard, but it has a well-maintained code base [35] and is claimed to be twice as fast as TLSH, which is why it is considered in the evaluation. Variants of ssdeep [25] have addressed the problem related to small fragment detection, so the ssdeep the algorithm can no longer be excluded on this basis. For reference, the mrsh-hbft algorithm is very slow compared with the other algorithms; it would be much too slow for network traffic analysis.

3.2   Evaluation Methodology

Since it is difficult to evaluate algorithm performance in a real network with high precision, a controlled evaluation environment was created. The environment incorporated a network interface that listened to traffic coming in from a virtual server and destined to a virtual client. Only unencrypted TCP packets were transmitted because they led to the HTTP and FTP network packets that were dominant in file transmission. As mentioned earlier, dealing with encrypted traffic is outside the scope of this research.

In the experiments, each packet was stripped off its header and its payload was compared against the target hash. Since none of the selected algorithms had a built-in live filtering function for network traffic, this feature was added to each algorithm. The loopback interface was used to pass unencrypted TCP packets to each algorithm, which reported whether or not the payloads were recognized.

Most approximate matching algorithms only confirm how many chunks of a target hash are found in input data. Therefore, if the hashes of multiple files are compared against a single file, an algorithm can only tell if a file is present, not which file from among all the others. When filtering packet payloads, this means that an algorithm can tell if a packet contains a hash, but not which file is contained when multiple files are identified. This means that packet payloads matched by an algorithm have to be verified later.

When a packet payload is matched as belonging to a file, the payload is concatenated with all the other matched payloads and fed back to the algorithm after passing through the network filtering component. Since the order in which the randomly-chosen files were sent and their sizes were recorded, a second comparison of the packet payloads against the corpus of all possible files was performed to reveal whether each algorithm correctly identified the files transferred over the network interface. The only alternative would be to modify each algorithm to concretely identify one file out of many, but this could degrade the performance of the algorithm in a manner that was not intended by its developers.

Each algorithm essentially has a filter of the entire *t5-corpus* that it compares against each packet and, thus, identifies the files that were transmitted. A `curl` job on the client side pulled random files with adjustable throughputs over the loopback interface. The files were recorded along with whether or not they were recognized by the algorithm. The bandwidth of the algorithm was limited by its ability to recognize a payload before the next payload arrived. If the algorithm was still handling a payload when a new payload arrived on the network interface, then the new payload was lost. Thus, the algorithm could not recognize all the files that passed through. As the throughput increased, the file detection rate decreased.

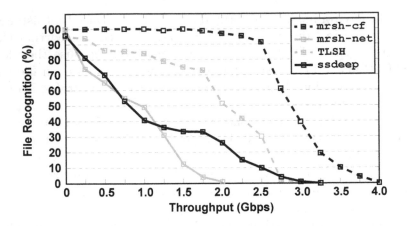

Figure 1. Initial throughput evaluations using the *t5-corpus*.

Figure 1 presents the results of initial evaluations of all the algorithms. The evaluations were performed on a workstation with a 3.5 GHz Intel Core i7-7500U mobile CPU (single-threaded). The throughput was incremented in steps of 250 Mbps and runs were performed ten times for each algorithm. The results show the efficacy of the algorithms at recognizing the randomized *t5-corpus* transferred over the loopback interface at various throughputs. Note that the non-optimized evaluations compared packet payloads against the algorithm hashes. The deviations over several runs were negligible. The evaluations demonstrate the ability of the algorithms to filter files correctly. Only true positives were considered, no false positives. Note that the filter performance of most of the algorithms decreased rapidly as the throughput increased.

The following statements can be made about the algorithms:

- The TLSH algorithm can detect finer-grained similarities compared with ssdeep, but it is more focused on differentiation than identification. TLSH returns a score for each comparison, where zero indicates a perfect match and higher scores indicate that the files are less similar. Only scores below 300 were accepted because they were either true or false positives. If the scoring threshold is set to a number below 300, there is a risk of not recognizing all the files. The 300 score appears to be the right threshold; if distances beyond this score are accepted, then the false positive rate becomes too high and would include true positives by accident. TLSH, which is best suited to evaluate the degree of similarity between two files, is used by VirusTotal to determine the similarities between malware variants. When TLSH was used to identify a single file in the *t5-corpus* (with no time constraints and counting all files below 300 as positives), up to 100 false positives were recorded depending on the filetype. Not visible in this evaluation, but nevertheless noteworthy, is that TLSH had a high false positive rate.

- The **ssdeep** algorithm was only able to detect file fragments containing 25-50% of the original files [27]. This issue may have been addressed in later versions of the algorithm. The evaluation used the standard version of **ssdeep** [35], which struggled to identify small fragments. With regard to the visibility of files on a per-payload basis, this means that smaller payloads cannot be matched correctly. The results show that **ssdeep** can filter some files at high throughputs that the other algorithms cannot handle. This is seen in Figure 1 as the long flattening curve that reaches 3.25 Gbps on the x-axis. The inability of **ssdeep** to match small fragments contained in payloads leads to poor detection rates.

- The evaluations reveal that **mrsh-cf** is by far the most powerful algorithm. Its superior cuckoo filter yields much better results than its predecessor **mrsh-net**. Further improvements were obtained because the **mrsh-cf** version used in this research incorporated an improved cuckoo filter compared with the original version. In the original **mrsh-net** evaluation [4], input packets had a uniform size of 18 chunks, and if 12 out of 18 chunks were found by the filter, then the packet was considered a match. As explained in Section 2.1, the evaluations conducted in this research employed variable-size payloads to better simulate real-world network scenarios.

- The **mrsh-net** algorithm was the only one originally intended for a network traffic matching scenario. The high discrepancy between the maximum throughput of 650 Mbps with a 99.6% true positive rate reported by the algorithm developers [4] and the results obtained in this research is probably due to the network scenarios being very different. Specifically, the developers of **mrsh-net** evaluated it in a simulated network scenario in which the *t5-corpus* was divided into equal-sized chunks (average TCP payload size of 1,460 bytes) that were input to the algorithm. The algorithm performed rather poorly in this evaluation due to the reduced runtime efficiency in a real network environment.

A closer investigation was conducted on the two best performing algorithms, TLSH and **mrsh-cf**. The interesting result is that the certainty with which the algorithms identified positives differed considerably. The

Table 4. mrsh-cf error rates for various throughputs.

Measure	Throughput (Mbps)				
	500	1,000	1,500	2,000	2,500
True Positives	2,000	2,000	2,000	1,997	1,891
False Positives	1,904	1,764	1,703	1,205	1,002
Average False Positive Size (%)	2.41	3.77	6.03	8.25	9.00
Average True Positive Size (%)	83.0	79.5	64.9	59.5	49.5

mrsh-cf algorithm found most chunks to be true positives and the false positives were mostly due to less than 10% matches of a file. Further filtering of false positives could be accomplished by applying a threshold. For example, if a file is identified with less then 10% of its chunks, then it is most likely a false positive and may be ignored. However, this applies only when a file is transferred in its entirety. When portions of a file are transferred, this heuristic might lead to false negatives.

In contrast, the TLSH algorithm identified true and false positives with the same certainty when the threshold for a positive was set below 300 (of 1,000). Positives were all identified as being 80% identical to the input file. Unlike, the mrsh-cf algorithm, a true positive is indistinguishable from among all the positives found by TLSH. This makes mrsh-cf the preferred algorithm for detecting unmanipulated files.

4.   Experimental Results and Optimizations

In the evaluations, the mrsh-cf algorithm consistently achieved the highest detection rates of all algorithms. Using mrsh-cf as an exemplar, experiments were conducted to demonstrate the significance of the false positive rate and how an algorithm can be optimized using a heuristic. In the case of an approximate matching algorithm, a false positive corresponds to a file that was falsely matched. If an approximate matching algorithm is used to blacklist files in network traffic, then a high false positive rate can significantly reduce its utility.

Table 4 shows the number of false positives obtained for various throughputs. The main observation is that the number of false positives decreases with higher throughputs. This is because there are too many packets at higher throughputs and the mrsh-cf algorithm is unable to keep up and match all the data against the filter. Nevertheless, the ratio of true positives to false positives remains around 1/3.

The mrsh-cf algorithm is impractical as a packet filter because it drops too much traffic. In fact, it drops a packet as soon as it recognizes a file in its filter; this is the most aggressive form of payload filtering. However, the experiments show that this can be adjusted to increase accuracy while decreasing speed. The average false positive size shows how much of a false positive file was recognized on average. The algorithm can be adjusted based on this value to reduce its false positive rate, for example, by requiring at least $x\%$ of a file to be found before it is considered a match. With this threshold, the algorithm can be adjusted to mark files as positive only if a certain percentage of the target hash has been recognized. A suitable threshold enabled the false positive rate of the algorithm to be reduced by approximately 90%.

As mentioned above, in its fastest "mode," `mrsh-cf` can only determine the presence of a filter file in a payload. It can reveal how many chunks in the filter match the input payload, but not the exact file that matched because the filter does not hold this information. Only through exclusion – comparing file after file with the payload — can the algorithm narrow down the exact matching file.

The algorithm detection rate could be improved by chaining multiple instances of the algorithm that run in different modes. Harichandran et al. [23] have hinted at this possibility. First, a rapid (rough) preselection is performed using an instance of the algorithm that compares payloads against the filter. Next, all the positives are compared by individual instances of the algorithm that compare them against every file in the filter. As Breitinger and Baggili [4] have remarked, hashing algorithms can greatly benefit from parallelization by using multiple threads to compare hashes. As discussed below, the behavior of each of the selected algorithms was evaluated under multi-threading. The results for `mrsh-cf`, the fastest adapted algorithm, are presented.

Figure 2 shows that delegating the comparison task to two threads that do not require synchronization increased the throughput at which all the files were detected correctly by 100%. Specifically, the single threaded version attributed files 100% correctly at a maximum throughput of 1.5 Gbps whereas adding a second thread increased the throughput at which all files were detected correctly to 3 Gbps. Triple threading increased the throughput even further to 4 Gbps with a recognition rate of 97%. It can be assumed that adding more threads, which is not a problem with modern hardware, would increase the throughput even more while keeping the recognition rate constant.

However, the robustness of the algorithm may become an issue. At this time, noise and entangled files encountered in live traffic scenarios negatively impact robustness. Breitinger and Baggili [4] have demonstrated that pre-sorting the files using common substring filtering and

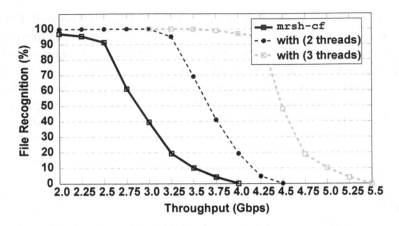

Figure 2.   Throughput evaluations using the *t5-corpus* (threaded and normal).

measures of anti-randomness greatly benefit the application of the algorithm in live traffic scenarios.

5.   Conclusions

This research has investigated the ability of prominent approximate matching algorithms to detect files in network traffic for purposes of data loss protection. In modern gigabit environments, network-focused data loss prevention solutions must be fast and reliable. For the first time, an effort has been made to adjust approximate matching algorithms to reliably recognize files at throughputs in the gigabit range. Optimizations have been introduced to render the algorithms more viable for live traffic detection. Indeed, the best algorithm, mrsh-cf, detects 97% of all files at a throughput of 4 Gbps in an idealized scenario. Open-source technology has been employed throughout this research and the algorithms are available to the digital forensics community at github.com/dasec/approx-network-traffic.

Future research will investigate techniques to reduce the false positive rates encountered when applying approximate matching algorithms in real-world network environments. The potential of matching several packets simultaneously to increase precision and incorporating multithreading to enhance file detection and throughput will be examined and the results integrated in the algorithms. Another problem involves reducing the bottleneck during payload inspection. Future work will also explore other promising approximate matching algorithms for file matching in network traffic; for example, Charyyev and Gunes [14] have shown that Nilsimsa hashes of Internet of Things traffic can be used as means for identification. A compelling problem is to perform holistic filtering of network traffic that would enable entire sets of network communications to be filtered by approximate matching techniques and matched by content and/or origin. This would address the challenges that encryption imposes on visibility in modern networks while refraining from TLS inspection.

Acknowledgement

This research was supported by the German Federal Ministry of Education and Research under Forschung an Fachhochschulen (Contract no. 13FH019IB6) and the Hessian Ministry of Higher Education, Research, Science and the Arts via their joint support of the National Research Center for Applied Cybersecurity ATHENE.

References

[1] S. Alneyadi, E. Sithirasenan and V. Muthukkumarasamy, A survey of data leakage prevention systems, *Journal of Network and Computer Applications*, vol. 62, pp. 137–152, 2016.

[2] H. Baier and F. Breitinger, Security aspects of piecewise hashing in computer forensics, *Proceedings of the Sixth International Conference on IT Security Incident Management and IT Forensics*, pp. 21–36, 2011.

[3] F. Breitinger, K. Astebol, H. Baier and C. Busch, mvHash-B – A new approach for similarity-preserving hashing, *Proceedings of the Seventh International Conference on IT Security Incident Management and IT Forensics*, pp. 33–44, 2013.

[4] F. Breitinger and I. Baggili, File detection in network traffic using approximate matching, *Journal of Digital Forensics, Security and Law*, vol. 9(2), pp. 23–36, 2014.

[5] F. Breitinger and H. Baier, A fuzzy hashing approach based on random sequences and Hamming distance, *Proceedings of the Annual ADFSL Conference on Digital Forensics, Security and Law*, pp. 89–100, 2012.

[6] F. Breitinger and H. Baier, Similarity-preserving hashing: Eligible properties and a new algorithm MRSH-v2, in *Digital Forensics and Cyber Crime*, M. Rogers and K. Seigfried-Spellar (Eds.), Springer, Berlin Heidelberg, Germany, pp. 167–182, 2013.

[7] F. Breitinger, H. Baier and J. Beckingham, Security and implementation analysis of the similarity digest sdhash, *Proceedings of the First International Baltic Conference on Network Security and Forensics*, 2012.

[8] F. Breitinger, B. Guttman, M. McCarrin, V. Roussev and D. White, Approximate Matching: Definition and Terminology, NIST Special Publication 800-168, National Institute of Standards and Technologies, Gaithersburg, Maryland, 2014.

[9] F. Breitinger, H. Liu, C. Winter, H. Baier, A. Rybalchenko and M. Steinebach, Towards a process model for hash functions in digital forensics, in *Digital Forensics and Cyber Crime*, P. Gladyshev, A. Marrington and I. Baggili (Eds.), Springer, Cham, Switzerland, pp. 170–186, 2014.

[10] F. Breitinger and K. Petrov, Reducing the time required for hashing operations, in *Advances in Digital Forensics IX*, G. Peterson and S. Shenoi (Eds.), Springer, Heidelberg, Germany, pp. 101–117, 2013.

[11] A. Breslow and N. Jayasena, Morton filters: Fast, compressed sparse cuckoo filters, *The VLDB Journal*, vol. 29(2-3), pp. 731–754, 2020.

[12] D. Chang, M. Ghosh, S. Sanadhya, M. Singh and D. White, FbHash: A new similarity hashing scheme for digital forensics, *Digital Investigation*, vol. 29(S), pp. S113–S123, 2019.

[13] D. Chang, S. Sanadhya and M. Singh, Security analysis of MVhash-B similarity hashing, *Journal of Digital Forensics, Security and Law*, vol. 11(2), pp. 22–34, 2016.

[14] B. Charyyev and M. Gunes, IoT traffic flow identification using locality-sensitive hashes, *Proceedings of the IEEE International Conference on Communications*, 2020.

[15] E. Damiani, S. De Capitani di Vimercati, S. Paraboschi and P. Samarati, An open digest-based technique for spam detection, *Proceedings of the ICSA Seventeenth International Conference on Parallel and Distributed Computing Systems*, pp. 559–564, 2004.

[16] Editorial Team, Our work with the DNC: Setting the record straight, *CrowdStrike Blog*, June 5, 2020.

[17] B. Fan, D. Andersen, M. Kaminsky and M. Mitzenmacher, Cuckoo filter: Practically better than Bloom, *Proceedings of the Tenth ACM International Conference on Emerging Networking Experiments and Technologies*, pp. 75–88, 2014.

[18] S. Garfinkel, P. Farrell, V. Roussev and G. Dinolt, Bringing science to digital forensics with standardized forensic corpora, *Digital Investigation*, vol. 6(S), pp. S2–S11, 2009.

[19] S. Gatlan, Software AG, IT giant, hit with $23 million ransom by Clop ransomware, *BleepingComputer*, October 9, 2020.

[20] T. Graf and D. Lemire, XOR filters: Faster and smaller than Bloom and cuckoo filters, *ACM Journal of Experimental Algorithmics*, vol. 25(1), article no. 5, 2020.

[21] V. Gupta and F. Breitinger, How cuckoo filters can improve existing approximate matching techniques, in *Digital Forensics and Cyber Crime*, J. James and F. Breitinger (Eds.), Springer, Cham, Switzerland, pp. 39–52, 2015.

[22] N. Harbour, dcfldd version 1.3.4-1 (dcfldd.sourceforge.net), 2006.

[23] V. Harichandran, F. Breitinger and I. Baggili, Bytewise approximate matching: The good, the bad and the unknown, *Journal of Digital Forensics, Security and Law*, vol. 11(2), pp. 59–78, 2016.

[24] J. Kornblum, Identifying almost identical files using context-triggered piecewise hashing, *Digital Investigation*, vol. 3(S), pp. 91–97, 2006.

[25] V. Martinez, F. Hernandez-Alvarez and L. Encinas, An improved bytewise approximate matching algorithm suitable for files of dissimilar sizes, *Mathematics*, vol. 8(4), article no. 503, 2020.

[26] J. Oliver, C. Cheng and Y. Chen, TLSH – A locality-sensitive hash, *Proceedings of the Fourth Cybercrime and Trustworthy Computing Workshop*, pp. 7–13, 2013.

[27] A. Lee and T. Atkison, A comparison of fuzzy hashes: Evaluation, guidelines and future suggestions, *Proceedings of the ACM South-East Conference*, pp. 18–25, 2017.

[28] D. Lillis, F. Breitinger and M. Scanlon, Expediting MRSH-v2 approximate matching with hierarchical Bloom filter trees, in *Digital Forensics and Cyber Crime*, P. Matousek and M. Schmiedecker (Eds.), Springer, Cham, Switzerland, pp. 144–157, 2018.

[29] F. Pagani, M. Dell'Amico and D. Balzarotti, Beyond precision and recall: Understanding uses (and misuses) of similarity hashes in binary analysis, *Proceedings of the Eighth ACM Conference on Data and Application Security and Privacy*, pp. 354–365, 2018.

[30] V. Roussev, Building a better similarity trap with statistically-improbable features, *Proceedings of the Forty-Second Hawaii International Conference on System Sciences*, 2009.

[31] V. Roussev, Data fingerprinting with similarity digests, in *Advances in Digital Forensics VI*, K. Chow and S. Shenoi (Eds.), Springer, Heidelberg, Germany, pp. 207–226, 2010.

[32] V. Roussev, An evaluation of forensic similarity hashes, *Digital Investigation*, vol. 8(S), pp. S34–S41, 2011.

[33] V. Roussev, G. Richard and L. Marziale, Multi-resolution similarity hashing, *Digital Investigation*, vol. 4(S), pp. S105–S113, 2007.

[34] C. Sadowski and G. Levin, SimHash: Hash-Based Similarity Detection, Technical Report, Department of Computer Science, University of California Santa Cruz, Santa Cruz, California, 2007.

[35] `ssdeep` Project, `ssdeep` – Fuzzy Hashing Program, GitHub (`ssdeep-project.github.io/ssdeep`), April 11, 2018.

III

ADVANCED FORENSIC TECHNIQUES

Chapter 6

LEVERAGING USB POWER DELIVERY IMPLEMENTATIONS FOR DIGITAL FORENSIC ACQUISITION

Gunnar Alendal, Stefan Axelsson and Geir Olav Dyrkolbotn

Abstract   Modern consumer devices present major challenges in digital forensic investigations due to security mechanisms that protect user data. The entire physical attack surface of a seized device such as a mobile phone must be considered in an effort to acquire data of forensic value.

Several USB protocols have been introduced in recent years, including Power Delivery, which enables negotiations of power delivery to or from attached devices. A key feature is that the protocol is handled by dedicated hardware that is beyond the control of the device operating systems. This self-contained design is a security liability with its own attack surface and undocumented trust relationships with other peripherals and the main system-on-chips.

This chapter presents a methodology for vulnerability discovery in USB Power Delivery implementations for Apple devices. The protocol and Apple-specific communications are reverse engineered, along with the firmware of the dedicated USB Power Delivery hardware. The investigation of the attack surface and potential security vulnerabilities can facilitate data acquisition in digital forensic investigations.

Keywords: Digital forensic acquisition, mobile device security, USB Power Delivery

1.   Introduction

Law enforcement has special opportunities to leverage security vulnerabilities in digital forensic acquisition. Since law enforcement can seize devices, any exposed physical interface on the devices is a potential attack vector for data acquisition.

Exposed interfaces on mobile devices that can be leveraged without physically opening the devices include SIM card slots, SD card slots, audio jacks and USB connectors. Interfaces that are exposed by phys-

© IFIP International Federation for Information Processing 2021
Published by Springer Nature Switzerland AG 2021
G. Peterson and S. Shenoi (Eds.): Advances in Digital Forensics XVII, IFIP AICT 612, pp. 111–133, 2021.
https://doi.org/10.1007/978-3-030-88381-2_6

ically opening devices are UART, JTAG and essentially any on-board peripherals that can be manipulated or replaced. The internal interfaces are often less accessible because opening a device like a mobile phone can be cumbersome and risky. A mobile phone is often glued shut and attempting to open it could disrupt normal operations, especially if the device must be powered on to exploit a specific vulnerability. This situation can occur if a phone is seized after the user has unlocked the device at least once since the device was powered on (i.e., after-first-unlock state), where user keys tied to user credentials are unlocked and more user data is potentially available. Thus, security vulnerabilities exposed via externally-accessible interfaces are preferred over internal interfaces.

The USB connector is one of the most common external interfaces in modern personal computers and embedded devices. As a result, the security of USB protocols is important from a digital forensic perspective. Wang et al. [31] have discussed USB attack strategies on the functional and physical layers, and several vulnerable scenarios for USB connected devices. However, they did not explore the security of the USB Power Delivery protocol.

Tian et al. [26] have investigated USB security. They have examined the security features provided by the USB Type-C connector, specifically, authentication that is included in recent USB Power Delivery revisions. They formally verified the authentication and identified USB attack vectors, but do not discuss implementation details. Examining actual implementations for verification of USB Power Delivery as an attack vector is, therefore, interesting and timely.

USB Power Delivery is a feature in newer devices that is available externally over standard USB physical interfaces such as a USB Type-C connector [28]. It is available on many modern personal computers and mobile phones in the after-first-unlock (AFU) and before-first-unlock (BFU) states. USB Power Delivery enables connected devices to negotiate the optimal power delivery (voltage and current), where one device acts as the source and the other as the sink.

Because devices can choose to swap the source and sink roles, the USB Power Delivery protocol supports the negotiation of the direction of power flow. Additionally, the protocol supports the negotiations of multiple devices connected to a single power source as well as re-negotiations at any time if more power is required. The protocol specification allows direct current levels up to 5 A, corresponding to a maximum of 100 W at 20 V. Thus, even the cables need to communicate using the USB Power Delivery protocol to ensure that they support higher current levels. These cables are named "electronically marked cables" (EMCA) [29].

USB Power Delivery employs messages for communications [29]. Revision 2.0 of the protocol has control and data messages. Revision 3.0 specifies additional extended messages that support features such as firmware updates, battery information, manufacturer information and security messages. USB Power Delivery also supports a side-band channel for standard and non-standard vendor-specific communications.

Thus, the source, sink and cable can transmit and receive control, data and extended messages, as well as additional vendor-specific messages. The presence of original and additional features of the protocol raise the question whether the protocol is secure from a vulnerability perspective. The code implementing such a feature-rich protocol is large and complex, increasing the likelihood of faults, which include security vulnerabilities. Estimating the ratio of security vulnerabilities per line of code is difficult and cumbersome. Hatton [15] suggests a defect (bug) density of less than ten per thousand lines of code. Ozment and Schechter [19], who evaluated OpenBSD, suggest a vulnerability density three orders of magnitude less. Although the figures are not directly comparable, they indicate that more code increases the likelihood of security vulnerabilities.

The complexity of the USB Power Delivery protocol and its codebase increase the likelihood that software security vulnerabilities may have been introduced during design and implementation. The protocol is also implemented in dedicated hardware, which raises questions about the state and integrity of the chip as well as the trust relationships with the rest of the system and the system-on-chip (SoC). This could expose the implementation to "evil maid" attacks [25] that replace the firmware in a USB Power Delivery chip or simply replace the entire chip.

The basis for any vulnerability research is the design and implementation details. Before applying any vulnerability discovery techniques [5, 17], access to code in any form and testing tools are extremely beneficial. Static vulnerability analysis benefits greatly from access to design and code details whereas fuzzing [24] requires simulation testing methods and tools.

Since most USB Power Delivery implementations are proprietary, the availability of source code is limited. Therefore, evaluating and estimating the likelihood of faults and security vulnerabilities based on lines of code is difficult. Additionally, since the protocol is implemented in dedicated hardware with undocumented interfaces, the ability to analyze and evaluate security via testing is limited. Few tools are available to perform black box testing or fuzzing of the protocol [1]. Extracting firmware from a USB Power Delivery chip and analyzing the firmware

are also difficult tasks. But they are important because such proprietary, non-scrutinized code may have many unknown security vulnerabilities.

It appears that USB Power Delivery has potential vulnerabilities in the protocol [23, 33] and its implementations [5]. Other vulnerabilities may arise from hidden features [8], implicit trust relationships [6] and hardware exposures such as evil maid [25]. Clearly, the exploitation of USB Power Delivery should be investigated as a means to enable the forensic acquisition of data from devices that incorporate the hardware and implement the protocol.

This chapter presents a new methodology for evaluating the potential of USB Power Delivery as an attack vector. The focus on USB Power Delivery implementations can provide insights into vulnerabilities. Binary diffing [10] of firmware versions can reveal security patches that can be exploited. Indeed, this research is important to understanding and leveraging USB Power Delivery as an entirely new attack surface for forensic data acquisition.

2.   USB Power Delivery Protocol

The USB Power Delivery protocol specifications were released in 2012 as Revision 1.0 (version 1.0). The most recent specifications are provided in Revision 2.0 (version 1.3) and Revision 3.0 (version 2.0) [29]. USB Power Delivery offers a uniform method for devices to negotiate power supply configurations across vendors. The protocol is often used by devices with USB Type-C connectors. A USB Type-C connector has dedicated lines, CC1 and CC2, that enable USB Power Delivery communications between devices. However, a USB Type-C connector is not required to support USB Power Delivery; for example, Apple's proprietary Lightning Connector [12] supports USB Power Delivery. In fact, a cable with Apple Lightning and USB Type-C connectors could be used between an Apple device and any other USB Power Delivery enabled device to facilitate communications. The Apple Lightning and USB Type-C connectors are reversible, so the orientations of the connectors are not important.

The message-based USB Power Delivery protocol employs three types of messages: control, data and extended messages. Control messages are short messages that typically require no data exchange. Data messages contain data objects that are exchanged between devices. Extended messages, introduced in Revision 3.0, are data messages with larger payloads.

Figure 1 shows the format of a USB Power Delivery data message packet. The packet has a transport portion, comprising a preamble,

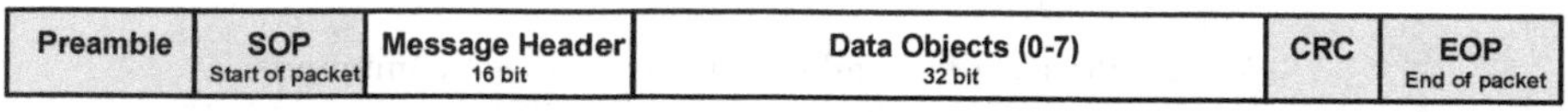

Preamble	SOP Start of packet	Message Header 16 bit	Data Objects (0-7) 32 bit	CRC	EOP End of packet

Figure 1. USB Power Delivery data message packet.

start of packet (SOP), cyclic redundancy check (CRC) and end of packet (EOP) fields, that encapsulates a message header and optional (up to seven) data objects. A data object has a fixed size of 32 bits, which allows for a maximum of 7×32 bits of data per message.

Table 1. Power Delivery protocol messages.

Control Messages Revs. 2.0 and 3.0	Data Messages Revs. 2.0 and 3.0	Extended Messages Rev. 3.0 only
GoodCRC	Source_Capabilities	Source_Capabilities_Extended
GotoMin	Request	Status
Accept	BIST	Get_Battery_Cap
Reject	Sink_Capabilities	Get_Battery_Status
Ping	Vendor_Defined	Battery_Capabilities
PS_RDY		Get_Manufacturer_Info
Get_Source_Cap	**Rev. 3.0 only**	Manufacturer_Info
Get_Sink_Cap	Enter_USB	Security_Request
DR_Swap	Battery_Status	Security_Response
PR_Swap	Alert	Firmware_Update_Request
VCONN_Swap	Get_Country_Info	Firmware_Update_Response
Wait		PPS_Status
Soft_Reset		Country_Info
		Sink_Capabilities_Extended
Rev. 3.0 only		Country_Codes
Data_Reset_Complete		
Not_Supported		
Get_Source_Cap_Extended		
Get_Status		
FR_Swap		
Get_PPS_Status		
Get_Country_Codes		
Get_Sink_Cap_Extended		
Data_Reset		

USB Power Delivery supports a wide range of standard messages to facilitate negotiations of power source configurations between devices. Table 1 lists the messages supported by Revision 2.0 (version 1.3) and/or Revision 3.0 (version 2.0). The backward compatibility means that protocol complexity increases in new revisions as new messages are added but existing messages are not eliminated.

Table 2.   Structured Vendor-Defined message commands.

Structured VDM Commands
Discover Identity
Discover SVIDs
Discover Modes
Enter Mode
Exit Mode
Attention
SVID-Specific Commands

Some of the standard messages in Table 1 have sub-types. For example, a Vendor-Defined message (VDM) can be structured or unstructured. Structured VDMs have commands defined in the standard (Table 2). Unstructured VDM commands are defined by vendors and are undocumented. Vendors are free to implement proprietary communications using unstructured VDMs as demonstrated in [1]. Enabling vendors to add messages over and above the standard messages results in increased complexity and firmware code size.

To avoid conflicts when implementing proprietary vendor messages, VDMs require a standard vendor ID (SVID) defined in the specification or a vendor ID (VID) to be part of the VDM header. A VID is a unique 16-bit identifier assigned by the USB Implementers Forum [30]. A vendor with a valid VID is free to implement any VDMs needed to operate its USB Power Delivery enabled devices. Apple devices commonly use the VID 0x05ac [13].

Connected devices negotiate power delivery via an explicit contract. Typically, this is initiated by the source device that sends a Source-Capabilities data message to which the sink replies with a GoodCRC message followed by a Request message (Figure 2). These responses inform the source that the sink is USB Power Delivery enabled and the highest protocol revision it supports. Specification revision information is included in the message header of the Request message. The highest specification revision supported by the sink corresponds to the highest specification revision supported by the source, which is indicated in the Source-Capabilities message. Thus, the connected devices know the revision and the message sets that are mutually supported.

3.   Research Methodology

The methodology for researching USB Power Delivery firmware involves information gathering, monitoring black box testing and simu-

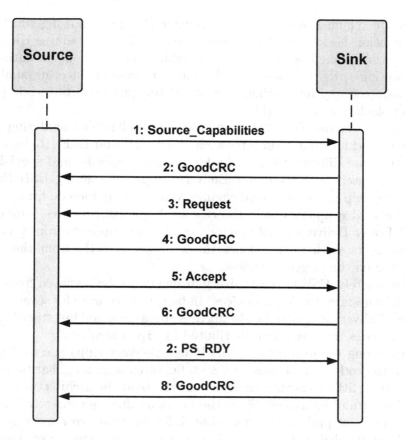

Figure 2. Generic, source-initiated explicit contract negotiation.

lation, and reverse engineering actual implementations (using binary code, documentation, source code, etc.). The individual methods often aid and overlap each other. For example, static reverse engineering of a binary is often assisted by monitoring and simulation. Even more powerful methods are instrumentation and debugging, which advance reverse engineering and vulnerability discovery.

Access to the source code of USB Power Delivery implementations (firmware) is difficult. This is because the source code is developed by the chip vendor and/or device vendor and are considered to be proprietary, business-confidential information. The documentation is also considered confidential and is kept in-house. As a result, the only option for researchers intending to study USB Power Delivery implementations is to extract and reverse engineer firmware.

Reverse engineering firmware involves the extraction of machine code (binary) for a specific chip and applying static and dynamic methods to produce human-readable assembly code [14, 18], following which decompilation is performed to obtain pseudo high-level source code [7]. Static reverse engineering analyzes machine code without executing or interacting with the code [14]. Dynamic reverse engineering analyzes machine code by interacting with and debugging executing code (e.g., using black box testing) [14].

Static and dynamic methods and tools are available for analyzing well-known machine code structures such as PE [21] and ELF [16] binaries, but they are difficult to come by for hardware-specific and specialized firmware used by USB Power Delivery chips. Obtaining USB Power Delivery chip firmware is challenging. Vendors may include firmware in regular device updates as in the case of Apple iOS updates. However, USB Power Delivery chips may not receive any updates from vendors, requiring researchers to extract the firmware directly from the chips soldered on the targeted devices.

Fortunately, USB Power Delivery firmware can be retrieved from general iOS updates for Apple devices. In fact, the firmware for several USB Power Delivery chips can be obtained by unpacking and investigating the iOS updates that are often distributed in `.ipsw` archives.

Analyzing the firmware of USB Power Delivery chips is complex because the code may be based on specific, often unknown, hardware. In particular, little to nothing may be known about the architecture, memory layout and interfaces. Since the firmware does not directly interact with users, helpful, human-readable, informational/error messages are rarely embedded in the code. This renders static reverse engineering very difficult, making dynamic reverse engineering the only feasible option.

Dynamic reverse engineering involves the execution and observation of the behavior of firmware. A simulation tool can be used to evaluate code execution by communicating with the USB Power Delivery interface and, consequently, the firmware code. This can be accomplished using a proprietary USB Power Delivery simulation device [1]. USB Power Delivery messages are sent to the device to assist with static reverse engineering. Specifically, responses to the messages are matched in a trial and error manner to identity the corresponding firmware code. However, reversing the firmware and resulting assembly code are still very tedious. Also, the reverse engineering results may not fully match the full feature set of the original source code. Nevertheless, the results would help understand unknown parts of the protocol such as VDMs. The (re)produced pseudo code from the (re)produced assembly code could be used to estimate of the lines of code in the original source code and, thus, the complexity of the implementation.

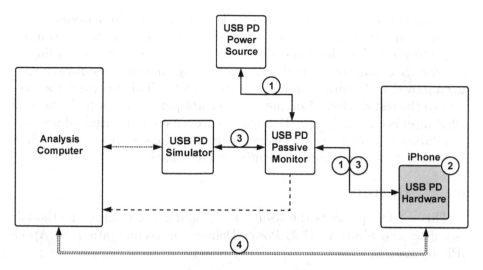

Figure 3. Experimental setup.

The production of pseudo code from firmware binaries from different vendors is challenging and difficult to generalize. Therefore, this research opted to pursue a full reverse engineering effort for targeted devices such as Apple iPhones to help generalize the results to a wider selection of devices in the future.

Figure 3 shows the experimental setup. It incorporates an analysis computer, target iPhone, USB Power Delivery monitor, USB Power Delivery simulator and stock USB Power Delivery power sources.

The general workflow is to first conduct information gathering from open sources. A generic USB PD passive monitoring tool (1 in Figure 3) is employed to observe device functionality (especially beyond the specified behavior) when the device is connected to other Apple devices and devices from other vendors. This provides early indications of proprietary vendor code and supports subsequent reverse engineering.

Passive monitoring of USB Power Delivery communications only covers the use of a subset of the protocol and optional vendor-specific messages. Therefore, the USB Power Delivery chip firmware must be extracted from the selected Apple device (2 in Figure 3) and static reverse engineering techniques employed to disassemble the firmware and reproduce the pseudo code via decompilation.

Next, dynamic reverse engineering techniques and trial-and-error probing of the running device with actual messages are performed (3 in Figure 3). These augment the static reverse engineering efforts to provide a better understanding of the firmware. After important components of the firmware, especially undocumented vendor-specific functionality, are understood, attempts are made to implement and simulate the components to verify that the Apple device behaves and responds according to the reverse engineering results. A jailbreaking solution, `checkra1n` [20], is employed to facilitate on-device experiments. This provides root access to the test devices. Communications are performed over the normal USB interface (4 in Figure 3). Dynamic reverse engineering efforts and simulation of USB Power Delivery communications are also useful when conducting injection tests to identify vulnerabilities.

4.   Results

This section presents the results of applying the research methodology discussed above to USB Power Delivery implementations in Apple iPhone models.

4.1   Information Gathering

The USB Power Delivery hardware in iPhone X, iPhone 8 and iPhone 8 Plus models appears to employ a Cypress CYPD2104 embedded microcontroller [32]. General datasheets and hardware design guides are available at the vendor site [9]. While the documentation provides useful insights, it was not possible to obtain complete documentation for the hardware, which would have provided useful information about the memory mapping of peripherals.

A Cypress CYPD2104 embedded microcontroller has a 48 MHz ARM Cortex-M0 CPU with 32 KB of flash storage for firmware, 4 KB SROM for booting and configuration, and 4 KB of SRAM. The I/O subsystem includes two serial communications blocks supporting I2C, SPI and UART, as well as several GPIOs. The interfaces are used to communicate with other peripherals such as an Apple system-on-chip. Another feature that is relevant to reverse engineering and vulnerability discovery is Serial Wire Debug (SWD) access. The access could be over a JTAG interface that would enable on-device debugging capabilities, very useful for reverse engineering and vulnerability discovery. However, no attempts were made to access the interface in this research.

4.2   Passive Monitoring

USB Power Delivery is not enabled on all Apple devices. Tests on Apple iPhones suggest that it is supported by Apple iPhone 8 and later models. Using a commercial USB Power Delivery analyzer [27] on a USB Power Delivery enabled power supply connected to an iPhone reveals if the device supports USB Power Delivery. A supported device also reveals its specification revision (and thus the supported messages). An attempt by a source to negotiate an explicit contract with an unsupported device fails to elicit a response.

Table 3 shows a summary of the messages exchanged during an explicit contract between a source non-Apple power supply (Revision 3.0) and sink iPhone 10,6 (iOS 13.2.2). Table 4 shows a summary of the messages exchanged during an explicit contract between a source Apple power supply (Revision 2.0) and sink iPhone 10,6. Note that the GoodCRC messages are removed.

The non-Apple power supply supports Revision 3.0 (Index 115 in Table 3 and the test iPhone responds with Revision 2.0 (Index 127 in Table 3). This identifies the latest revision supported by the test device.

Connecting an Apple power supply reveals additional, vendor-specific communications (Table 4). Note that the communications are summarized and the GoodCRC messages are removed. The undocumented Apple device communications start with the first unstructured VDM with Index 299. This conforms to the USB Power Delivery specifications, which state that proprietary communications should use unstructured VDMs. The communications are initiated when the Apple power supply asks the iPhone for its device ID. The iPhone device responds with Apple VID 0x05ac, following which the Apple power source initializes and starts the Apple-device-specific protocol. This protocol is dissected in Section 4.5.

4.3   Firmware Files

The USB Power Delivery firmware files were located in iOS updates [2]. These files are regularly released by Apple to update the iOS operating system and support on-board peripherals. The firmware files reside in an unpacked .ipsw file and are named USB-C_HPM,x.bin, where x varies based on the number of included firmware files.

The firmware files come in various versions for installation on device hardware (Section 4.1). Many of the files are of the same size with minor differences in their binaries. An important difference is that each has a different product ID (PID) that is reported by its firmware in response to a structured Discover Identity VDM (Table 2).

Table 5 lists the firmware files and their PIDs. The PID enables any connected device to identify the iPhone model via the USB Power Delivery protocol. An important observation is that the firmware files essentially have identical code across all the PIDs (and thus iPhone models) for a given iOS version. This means that any security issues discovered in firmware for a specific iPhone model would likely be present in

Table 3. Explicit contract between source non-Apple power supply (Rev. 3.0) and sink iPhone X (iOS 13.2.2).

Revision	Index	Time	Role	Message	Data
3.0	115	0:35.294.997	Source:DFP	[0]Source_Cap	A1 61 2C 91 01 0A 2C D1 02 00 F4 21 03 00 F4 C1 03 00 B1 04 00 45 41 06 00 83 B5 F1 BC
	124	0:35.296.426	Sink:UFP	[0]GoodCRC	41 00 BB 6C BB A8
2.0	127	0:35.297.707	Sink:UFP	[0]Request	42 10 2C B1 04 13 3D 9D 18 5D
	131	0:35.298.419	Source:DFP	[0]GoodCRC	61 01 8F 78 38 4A
2.0	134	0:35.301.522	Source:DFP	[1]Accept	63 03 21 7B 00 96
2.0	137	0:35.302.329	Sink:UFP	[1]GoodCRC	41 02 97 0D B5 46
2.0	140	0:35.412.687	Source:DFP	[2]PS_RDY	66 05 51 2A 14 02
	143	0:35.413.232	Sink:UFP	[2]GoodCRC	41 04 A2 A8 D6 AF

Table 4. Explicit contract between source Apple power supply (Rev. 2.0) and sink iPhone 10,6 (iOS 13.2.2).

Revision	Index	Time	Role	Message	Data
2.0	182	0:21.801.787	Source:DFP	[3]Source_Cap	61 17 F0 90 01 08 EA 21 1F CC
2.0	189	0:21.803.817	Sink:UFP	[0]Request	42 10 F0 C0 03 13 BC 0F E8 2B
2.0	197	0:21.805.451	Source:DFP	[4]Accept	63 09 3F 92 D5 76
2.0	203	0:21.834.345	Source:DFP	[5]PS_RDY	66 0B 56 07 AC E5
2.0	238	0:24.825.555	Source:DFP	[1]VDM:DiscIdentity	6F 13 01 80 00 FF 16 62 AB 1B
2.0	245	0:24.827.511	Sink:UFP	[2]VDM:DiscIdentity	4F 44 41 80 00 FF AC 05 00 54 00 00 00 00 00 21 7D 16 94 99 07 82
2.0	255	0:24.831.372	Source:DFP	[2]VDM:DiscSVID	6F 15 02 80 00 FF 58 38 5E 86
2.0	262	0:24.833.320	Sink:UFP	[3]VDM:DiscSVID	4F 26 42 80 00 FF 00 00 AC 05 F6 20 C2 26
2.0	270	0:24.837.322	Source:DFP	[3]VDM:DiscMode	6F 17 03 80 AC 05 BA E4 F8 1B
2.0	277	0:24.839.154	Sink:UFP	[4]VDM:DiscMode	4F 28 43 80 AC 05 02 00 00 00 72 AD 21 96
2.0	285	0:24.844.365	Source:DFP	[4]VDM:EnterMode	6F 19 04 81 AC 05 55 08 DD 38
2.0	292	0:24.846.477	Sink:UFP	[5]VDM:EnterMode	4F 1A 44 81 AC 05 8E 2F C5 E3
2.0	299	0:24.850.260	Source:DFP	[5]VDM:Unstructured	6F 1B 05 00 AC 05 E7 4D 56 1A
2.0	307	0:24.851.919	Sink:UFP	[6]VDM:Unstructured	4F 1C 15 00 AC 05 5E C3 C3 FF
2.0	315	0:24.853.357	Sink:UFP	[7]VDM:Attention	4F 3E 06 81 AC 05 02 01 AC 05 00 00 00 00 DC 69 C4 D9
2.0	324	0:24.856.927	Source:DFP	[6]VDM:Unstructured	6F 3D 02 01 AC 05 00 00 00 06 00 00 20 12 5E E4 81
2.0	333	0:24.858.774	Sink:UFP	[0]VDM:Unstructured	4F 10 12 00 AC 05 E6 16 E4 A7
2.0	340	0:24.860.209	Sink:UFP	[1]VDM:Attention	4F 32 06 81 AC 05 02 01 AC 05 04 00 00 00 70 47 1C CC
2.0	349	0:24.863.748	Source:DFP	[7]VDM:Unstructured	6F 3F 02 01 AC 05 04 00 00 00 00 02 08 00 D1 C4 AA B0
2.0	358	0:24.865.628	Sink:UFP	[2]VDM:Unstructured	4F 14 12 00 AC 05 26 B0 64 52

Table 5. Firmware files with their PIDs and test iPhone models.

File Name	sha1sum	PID	iPhone Model
USB-C_HPM,1.bin	83D9F3003DF9CC1915507BD090608AE0AA96CF5D	0x1654	
USB-C_HPM,2.bin	87BF3EEDA8C98081657F13B5B547924893EF0ED3	0x165d	
USB-C_HPM,3.bin	0314146681992F7A4FE8B0F69A7AB42CA159E76D	0x166c	iPhone 8
USB-C_HPM,4.bin	273A80375FE8FEC09D498221BE4729588818582F	0x167c	iPhone 8 Plus
USB-C_HPM,5.bin	ACC3DBE669E310E1FE3063725F5B436E807C83D94	0x167d	iPhone X
USB-C_HPM,6.bin	B7CE922CD8B3D0018E4861006A065C7C5FBD9D5B	0x1686	
USB-C_HPM,7.bin	916D096C8939F4E108A269A854198B95BD5A7BEE	0x1687	
USB-C_HPM,8.bin	4FC3EB5B1B0244C04F7D6BB6A917ACAAE9F7D56F	0x1688	

Table 6. USB-C_HPM,4.bin files in various iOS versions.

iOS Version	File Name	sha1sum(USB-C_HPM,4.bin)
13.4	iPhone_5.5_P3_13.4.17E255_Restore.ipsw	9767A86F62ABDC8C1046F4D807CC30DAB99A4693
13.3.1	iPhone_5.5_P3_13.3.1.17D50_Restore.ipsw	273A80375FE8FEC09D498221BE4729588818582F
13.3	iPhone_5.5_P3_13.3.17C54_Restore.ipsw	273A80375FE8FEC09D498221BE4729588818582F
13.2.3	iPhone_5.5_P3_13.2.3.17B111_Restore.ipsw	273A80375FE8FEC09D498221BE4729588818582F
13.2.2	iPhone_5.5_P3_13.2.2.17B102_Restore.ipsw	273A80375FE8FEC09D498221BE4729588818582F
13.1.3	iPhone_5.5_P3_13.1.3.17A878_Restore	273A80375FE8FEC09D498221BE4729588818582F
12.4.1	iPhone_5.5_P3_12.4.1.16G102.Restore	79CB8220D2C6F5917C1C11ED7B4BF733E3C9B1C8
12.3	iPhone_5.5_P3_12.3.16F156_Restore.ipsw	79CB8220D2C6F5917C1C11ED7B4BF733E3C9B1C8
12.2	iPhone_5.5_P3_12.2.16E227_Restore.ipsw	79CB8220D2C6F5917C1C11ED7B4BF733E3C9B1C8
12.0	iPhone_5.5_P3_12.0.16A366_Restore.ipsw	B374072044A97669A688A49E1723C55E9973A851
11.4.1	iPhone_5.5_P3_11.0.11.4.1.15G77_Restore.ipsw	1E20D8B4D54D6C092DA9B668A53AAAE81ABFA3EE
11.0	iPhone10,5-11.0.15A372.Restore.ipsw	1E20D8B4D54D6C092DA9B668A53AAAE81ABFA3EE

multiple models, increasing the applicability of a forensic data acquisition method. Reverse engineering results for one device model could be reused across models, saving time and resources.

Research revealed that the iOS 13.2.2 updates for iPhone 8, iPhone 8 Plus and iPhone X models included identical USB-C_HPM,x.bin files as shown in Table 5.

The common firmware codebase also supports binary diffing. Security patches discovered in two versions of a USB-C_HPM,x.bin file can be assumed to be present in the firmware of different iPhone models. This greatly reduces the resources needed to discover potential security patches across device models.

For a USB-C_HPM,x.bin file corresponding to an iPhone 8 Plus model (USB-C_HPM,4.bin), different iOS upgrades can be downloaded and analyzed to detect changes to the USB Power Delivery firmware. Comparing the sha1sum values for differences is adequate to indicate a patch because there does not appear to be any iOS-specific rebuilding or versioning changes embedded in the firmware, leaving it untouched between updates unless the actual USB Power Delivery firmware is updated.

Table 6 shows several iOS updates for the iPhone 8 Plus model and the corresponding sha1sum(USB-C_HPM,4.bin) values. The trend is that the USB Power Delivery firmware is updated rarely. In fact, one patch was retained in iOS versions 13.3.1 to 13.4.

4.4   Firmware Reverse Engineering

Since most of the firmware files corresponding to different PIDs have few differences, reverse engineering can focus on just one of the USB-C_-HPM,x.bin files in Table 5. The machine architecture is ARM little-endian and the code is in the ARM Thumb mode [4], which is a subset of the ARM instruction set that uses variable-length instructions, often for improved code density. The code is also what is often referred to as "bare metal" code, meaning it can execute without any other abstraction layer (e.g., underlying operating system). The code directly interacts with the Apple system-on-chip and other peripherals through an I/O subsystem, mapped at specific memory addresses. Without documentation about the underlying USB Power Delivery hardware, the addresses are hardware-specific and often unknown. Therefore, from a reverse engineering perspective, it is necessary to make assumptions when code uses such unknown, hard-coded (non-position independent) addresses.

Table 7 shows the results of disassembling and decompiling the most recent versions of the firmware file USB-C_HPM,4.bin listed in Table 6. The total numbers of lines of pseudo C code for the two files are slightly

Table 7.   `USB-C_HPM,4.bin` details for two iOS versions.

iOS Version	Revision	Functions	Code Bytes	OP Codes	Pseudo C Code Lines
13.4	2.0	248	18,310	8,419	6,247
13.2.2	2.0	249	18,598	8,552	6,206

more than 6,200. USB Power Delivery Revision 2.0 was previously confirmed via passive monitoring. Therefore, the code supports all the messages listed for Revision 2.0 (Table 1). Code that implements additional unstructured VDMs is also included. It is expected that the number of lines of code would grow significantly to support a later revision (e.g., Revision 3.0). This is because Revision 3.0 supports a large number of additional messages (Table 1).

As described in Section 4.2, all the Apple-specific messages were identified and reverse engineered. Therefore, all the unstructured VDMs supported by the firmware could be identified in the disassembled code and pseudo C code. In fact, all the Apple-specific unstructured VDMs are handled by the same `handler` function. This function processes user input and is, therefore, an attractive target for vulnerability analysis. Erroneous handling of data in any USB Power Delivery message is a potential attack vector that could lead to a compromise of the USB Power Delivery functionality.

The number of lines of pseudo C code is relatively small compared with larger source code trees [19]. Since the firmware is "bare metal" code, the code is less generic and more difficult to compare with other sources. Therefore, the likelihood of security vulnerabilities in the firmware is difficult to compare with other estimates. Nevertheless, the code is in a state that is amenable to the application of established vulnerability discovery techniques [5, 17, 24].

4.5   Apple Vendor-Defined Protocol

The undocumented VDMs in Table 4 indicate that a special protocol is used by Apple devices to exchange information. Two connected Apple devices engage in an explicit contract negotiation as seen in the messages with Index 182 through 203 in Table 4. After this, the Apple-enabled power source requests the identity of the Apple iPhone X sink via a Discover Identity VDM with Index 238. Since the sink responds with the known Apple VID `0x05ac` and PID `0x167d`, the two devices can engage in additional communications using messages with Index 255

Vendor ID (VID) Bit 31...16	VDM Type Bit 15	Vendor use Bit 14...0

Figure 4. Unstructured VDM header.

through 292. The next message (Index 299) from the source to the sink is the first unstructured VDM and the first fully vendor-specific message. Upon dissecting the raw data in this message, bytes [0:2] were determined to correspond to the USB Power Delivery message header (Figure 1), bytes [2:6] to the VDM header and bytes [6:10] to the message CRC. Further dissection of the VDM header bytes [6:10] (little-endian) revealed the VID 0x05ac, VDM Type 0 (unstructured message) and Vendor Use 0x5 (Figure 4). The unstructured VDM was determined to contain the expected Apple VID 0x05ac and an undocumented command 0x5. For each undocumented command, a handler function can be identified in the associated firmware file USB-C_HPM,4.bin and disassembled.

Further dissection of the communications in Table 4 focused on the Attention VDM with Index 315 sent by the sink to the source (i.e., the iPhone asks the Apple power source for information). The response from the source has Index 324. This is interesting because the iPhone only requests this type of information when it is connected to an Apple device. In fact, it turns out that the iPhone requests a range of data from the Apple power source (serial number, device name, manufacturer, etc.).

Root access to the iPhone was achieved using checkra1n [20]. This enabled the recovery of the data exchanged using the Apple-specific protocol. Next, the command ioreg -f -i -l -w0 > /tmp/ioreg.txt was used to obtain the content of the iPhone I/O Registry [3], which made it possible to interpret the exchanged data.

Table 8 shows example data exchanged between the Apple power supply source and the iPhone sink. Note that the communications are summarized and the GoodCRC messages are removed. The ASCII data C04650505D5GW85A8 at Index 401 was located in the Apple I/O Registry [3] as the Apple power device serial number. The data can be located using the command ioreg -f -i -l -w0 | grep C04650, which returned "SerialNumber"="C04650505D5GW85A8" and "SerialString" ="C04650505D5GW85A8".

By leveraging the handler functions in the firmware, it is possible to identify all the implemented vendor protocol messages and, thus, all the supported unstructured VDMs and messages required by the USB Power Delivery protocol. Control over all the supported messages coupled with the ability to communicate with the iPhone hardware facilitates the discovery and exploitation of security vulnerabilities. These include direct code execution on the iPhone hardware and poor input validation by peripherals/system-on-chip/kernel using the USB Power Delivery data (user input).

Table 8 shows an example of the Apple power supply sending its serial number. Because all the messages supported by the firmware (including undocumented VDMs) can be replicated, all the data exchanged by the undocumented protocol can be modified at will.

4.6   Firmware Modification and Rollback

Analysis reveals that the USB-C_HPM,x.bin firmware files are unsigned and are, therefore, neither verified at installation time nor run time. This is verified by modifying the PID in a USB-C_HPM,5.bin file (see Table 5) and flashing it to the appropriate iPhone test device. With the aid of the checkra1n [20] jailbreaking solution, the Apple USB Power Delivery firmware flash executable usbcfwflasher included in the iOS firmware update file may be used to flash the modified USB-C_HPM,5.bin file. This can be done to all checkra1n-supported Apple devices without requiring any user credentials.

The firmware modification is verified by monitoring a normal explicit contract with the additional Apple-specific VDM protocol between an Apple power supply and iPhone with the modified firmware. A successful firmware modification results in a different PID being returned from the iPhone in response to a structured Discover Identity VDM from the power supply.

Table 9 shows that the returned PID in the message with Index 83 is 0x1337 instead of the expected PID 0x167d in the message with Index 245 in Table 4. Note that the communications are summarized and the GoodCRC messages are removed. The PIDs are the 16-bit little-endian values at bytes [16:18] in both messages.

The result is that it is possible to fully modify the USB Power Delivery firmware. This includes the ability to perform a firmware rollback and install an older, potentially-vulnerable, firmware version on a patched device. Because this is a security vulnerability in itself, it is very useful for further vulnerability discovery because researchers can implement any test code to expose, for example, further propagation in an iPhone or side-channel attack scenarios.

Table 8. Data exchanged between source Apple power supply (Rev. 2.0) and sink iPhone 10,6 (iOS 13.2.2).

Revision	Index	Time	Role	Message	Data	ASCII
2.0	392	0:24.874.135	Sink:UFP	[5]VDM:Attention	4F 3A 06 81 AC 05 02 05 AC 05 30 00 00 00 F8 DF 19 1D	O:.....: ..O..... ..
2.0	401	0:24.877.949	Source:DFP	[1]VDM:Unstructured	6F 73 02 05 AC 05 30 00 00 00 43 30 34 36 35 30 35 30 35 44 35 47 57 38 35 41 38 00 00 00 55 8A 48 BE	os.....0. ..C04650 505D5GW8 5A8...U. H.

Table 9. Discover Identity VDMs between source Apple power supply (Rev. 2.0) and sink iPhone 10,6 (iOS 13.2.2).

Revision	Index	Time	Role	Message	Data
2.0	76	0:44.204.575	Source:DFP	[7]VDM:DiscIdentity	6F 1F 01 80 00 FF 17 8F 5B DE
2.0	83	0:44.206.298	Sink:UFP	[2]VDM:DiscIdentity	4F 44 41 80 00 FF AC 05 00 54 00 00 00 00 00 21 37 13 94 CA FB F8

5.   Conclusions

The methodology for analyzing USB Power Delivery implementations facilitates the discovery of security vulnerabilities for exploiting USB Power Delivery hardware to acquire data in digital forensic investigations. The ultimate goal is to leverage system privileges, potentially through a new set of security vulnerabilities that are identified using the hardware as a springboard. Examples include vulnerable implicit trust relationships, USB Power Delivery hardware components and system processes that parse data provided as inputs by Apple VDM commands. The step-by-step methodology, which is demonstrated to expose the implementation details of USB Power Delivery devices, is applicable to a wide range of USB Power Delivery implementations by diverse vendors.

The results of using the methodology on Apple iPhones can be summarized as follows. Gathering information about the USB Power Delivery hardware assists firmware reverse engineering, side-channel analysis and attack development. The ability to monitor USB Power Delivery messages facilitates the analysis of messages supported by the device and helps discern if a proprietary vendor protocol is employed. Sending and receiving arbitrary messages using a simulation tool advances black box testing, reverse engineering and exploitation.

Additionally, the reverse engineering of firmware to yield disassembled code and pseudo C code is very useful for manual and automated vulnerability analyses. Diffing tests can help reveal patches that can be checked for potential security vulnerabilities. Rollbacks of vulnerable firmware can be accomplished on jailbroken devices without requiring user credentials because firmware signatures and rollback protection mechanisms are not implemented. The lack of signatures also facilitates arbitrary modifications of firmware that expose USB Power Delivery to evil maid attacks.

Future research will attempt to discover additional vulnerabilities. It will also attempt to simulate and instrument/debug the extracted firmware, with the goal of advancing fuzzing techniques for vulnerability discovery. Other topics for future research include debugging test devices and chips via JTAG and conducting simulation via emulation and symbolic execution [11, 22, 24].

Acknowledgements

This research was supported by the IKTPLUSS Program of the Norwegian Research Council under R&D Project Ars Forensica Grant Agreement 248094/O70. Apple was notified about this research in advance of publication.

References

[1] G. Alendal, S. Axelsson and G. Dyrkolbotn, Exploiting vendor-defined messages in the USB Power Delivery protocol, in *Advances in Digital Forensics XV*, G. Peterson and S. Shenoi (Eds.), Springer, Cham, Switzerland, pp. 101–118, 2019.

[2] Apple, About iOS 13 Updates, Cupertino, California (`support.apple.com/en-us/HT210393`), 2021.

[3] Apple, The I/O Registry, Cupertino, California (`developer.apple.com/library/archive/documentation/DeviceDrivers/Conceptual/IOKitFundamentals/TheRegistry/TheRegistry.html`), 2021.

[4] ARM, The Thumb Instruction Set, ARM7TDMI Technical Reference Manual, Revision r4p1, Cambridge, United Kingdom (`infocenter.arm.com/help/index.jsp?topic=/com.arm.doc.ddi0210c/CACBCAAE.html`), 2004.

[5] A. Austin and L. Williams, One technique is not enough: A comparison of vulnerability discovery techniques, *Proceedings of the International Symposium on Empirical Software Engineering and Measurement*, pp. 97–106, 2011.

[6] G. Beniamini, Over The Air: Exploiting Broadcom's Wi-Fi Stack (Part 2), Project Zero Team, Google, Mountain View, California (`googleprojectzero.blogspot.com/2017/04/over-air-exploiting-broadcoms-wi-fi_11.html`), 2017.

[7] G. Chen, Z. Qi, S. Huang, K. Ni, Y. Zheng, W. Binder and H. Guan, A refined decompiler to generate C code with high readability, *Software: Practice and Experience*, vol. 43(11), pp. 1337–1358, 2013.

[8] W. Chen and J. Bhadra, Striking a balance between SoC security and debug requirements, *Proceedings of the Twenty-Ninth IEEE International System-on-Chip Conference*, pp. 368–373, 2016.

[9] Cypress Semiconductor, CYPD2104-20FNXIT, San Jose, California (`www.cypress.com/part/cypd2104-20fnxit`), 2018.

[10] Y. Duan, X. Li, J. Wang and H. Yin, DeepBinDiff: Learning program-wide code representations for binary diffing, *Proceedings of the Twenty-Seventh Annual Network and Distributed System Security Symposium*, 2020.

[11] D. Engler and D. Dunbar, Under-constrained execution: Making automatic code destruction easy and scalable, *Proceedings of the ACM/SIGSOFT International Symposium on Software Testing and Analysis*, pp. 1–4, 2007.

[12] A. Golko, E. Jol, M. Schmidt and J. Terlizzi, Dual Orientation Connector with External Contacts and Conductive Frame, U.S. Patent No. 0115821 A1, May 9, 2013.

[13] S. Gowdy, The USB ID Repository (`www.linux-usb.org/usb-ids.html`) 2021.

[14] A. Harper, D. Regolado, R. Linn, S. Sims, B. Spasojevik, L. Martinez, M. Baucom, C. Eagle and S. Harris, *Gray Hat Hacking: The Ethical Hacker's Handbook*, McGraw-Hill Education, New York, 2018.

[15] L. Hatton, Re-examining the fault density component size connection, *IEEE Software*, vol. 14(2), pp. 89–97, 1997.

[16] Y. Li and J. Yan, ELF-based computer virus prevention technologies, *Proceedings of the Second International Conference on Information Computing and Applications*, pp. 621–628, 2011.

[17] B. Liu, L. Shi, Z. Cai and M. Li, Software vulnerability discovery techniques: A survey, *Proceedings of the Fourth International Conference on Multimedia Information Networking and Security*, pp. 152–156, 2012.

[18] A. Maurushat, *Disclosure of Security Vulnerabilities: Legal and Ethical Issues*, Springer, London, United Kingdom, 2013.

[19] A. Ozment and S. Schechter, Milk or wine: Does software security improve with age? *Proceedings of the Fifteenth USENIX Security Symposium*, 2006.

[20] A. Panhuyzen, `checkra1n` (`checkra.in`), 2021.

[21] M. Pietrek, Peering Inside the PE: A Tour of the Win32 Portable Executable File Format (`bytepointer.com/resources/pietrek_peering_inside_pe.htm`), 1994.

[22] E. Schwartz, T. Avgerinos and D. Brumley, All you ever wanted to know about dynamic taint analysis and forward symbolic execution (but might have been afraid to ask), *Proceedings of the IEEE Symposium on Security and Privacy*, pp. 317–331, 2010.

[23] A. Sosnovich, O. Grumberg and G. Nakibly, Finding security vulnerabilities in a network protocol using parameterized systems, *Proceedings of the Twenty-Fifth International Conference on Computer Aided Verification*, pp. 724–739, 2013.

[24] N. Stephens, J. Grosen, C. Salls, A. Dutcher, R. Wang, J. Corbetta, Y. Shoshitaishvili, C. Kruegel and G. Vigna, Driller: Augmenting fuzzing through selective symbolic execution, *Proceedings of the Twenty-Third Annual Network and Distributed System Security Symposium*, 2016.

[25] A. Tereshkin, Evil maid goes after PGP whole disk encryption, keynote lecture presented at the *Third International Conference on Security of Information and Networks*, 2010.

[26] J. Tian, N. Scaife, D. Kumar, M. Bailey, A. Bates and K. Butler, SoK: "Plug & pray" today – Understanding USB insecurity in versions 1 through C, *Proceedings of the IEEE Symposium on Security and Privacy*, pp. 1032–1047, 2018.

[27] Total Phase, USB Power Delivery Analyzer, Sunnyvale, California (`www.totalphase.com/products/usb-power-delivery-ana lyzer`), 2021.

[28] USB Implementers Forum, Universal Serial Bus Type-C Cable and Connector Specification, Release 2.0, Beaverton, Oregon (`www.usb.org/sites/default/files/USBType-CSpecR2.0-Augu st2019.pdf`), 2019.

[29] USB Implementers Forum, USB Power Delivery, Beaverton, Oregon (`www.usb.org/document-library/usb-power-delivery`), 2019.

[30] USB Implementers Forum, Getting a Vendor ID, Beaverton, Oregon (`www.usb.org/getting-vendor-id`), 2020.

[31] Z. Wang and A. Stavrou, Exploiting smartphone USB connectivity for fun and profit, *Proceedings of the Twenty-Sixth Annual Computer Security Applications Conference*, pp. 357–366, 2010.

[32] D. Yang, S. Wegner and J. Morrison, Apple iPhone X Teardown, TechInsights, Ottawa, Canada (`www.techinsights.com/blog/apple-iphone-x-teardown`), 2017.

[33] F. Yang and S. Manoharan, A security analysis of the OAuth protocol, *Proceedings of the IEEE Pacific Rim Conference on Communications, Computers and Signal Processing*, pp. 271–276, 2013.

Chapter 7

DETECTING MALICIOUS PDF DOCUMENTS USING SEMI-SUPERVISED MACHINE LEARNING

Jianguo Jiang, Nan Song, Min Yu, Kam-Pui Chow, Gang Li, Chao Liu and Weiqing Huang

Abstract   Portable Document Format (PDF) documents are often used as carriers of malicious code that launch attacks or steal personal information. Traditional manual and supervised-learning-based detection methods rely heavily on labeled samples of malicious documents. But this is problematic because very few labeled malicious samples are available in real-world scenarios.

This chapter presents a semi-supervised machine learning method for detecting malicious PDF documents. It extracts structural features as well as statistical features based on entropy sequences using the wavelet energy spectrum. A random sub-sampling strategy is employed to train multiple sub-classifiers. Each classifier is independent, which enhances the generalization capability during detection. The semi-supervised learning method enables labeled as well as unlabeled samples to be used to classify malicious and benign PDF documents. Experimental results demonstrate that the method yields an accuracy of 94% despite using training data with just 11% labeled malicious samples.

Keywords: Malicious PDF documents, machine learning, semi-supervised learning

1.   Introduction

Portable Document Format (PDF) documents are immensely popular due to their platform-independence and convenience, but this has led to their nefarious use as carriers of malicious code. Since the first PDF vulnerability was revealed in 2008 [2], the exploitation of PDF documents has proliferated. Phishing email with malicious PDF document attachments are often used to attack government organizations, companies and

© IFIP International Federation for Information Processing 2021

Published by Springer Nature Switzerland AG 2021

G. Peterson and S. Shenoi (Eds.): Advances in Digital Forensics XVII, IFIP AICT 612, pp. 135–155, 2021.

https://doi.org/10.1007/978-3-030-88381-2_7

individuals. FireEye [6] reports that advanced persistent threat groups such as APT39 often leverage malicious PDF documents.

A promising approach is to detect malicious documents in hosts and network traffic. Anti-virus systems employ signatures and heuristics for malicious document detection [21], but they are ineffective against new attacks and polymorphic attacks. Liu et al. [11] have detected malicious documents using statistical indicators such as document entropy, but have overlooked document structure information. Lin and Pao [10] employ content features for malicious PDF document detection; they extract JavaScript code, metadata and function keywords, and use a support vector machine to classify PDF documents as malicious. Srndic and Laskov [18] have extracted hierarchical path information in PDF document structures and engage decision tree and support vector machine classification algorithms to detect malicious documents.

Other detection approaches analyze the structural compositions and logical structures of PDF documents at runtime [12, 13, 17]. The Hidost [19] and SFEM [4] methods rely on structural features. Li et al. [9] consider structural features and JavaScript code and employ a feature extractor to detect malicious PDF documents. Yu et al. [26] use a model based on two layers of abstraction to detect malicious documents. Other recent work focuses on encryption [16] and adversarial PDF document samples [25]. In contrast, Xu and Kim [24] detect malicious PDF documents by opening them in a sandbox and monitoring their behavior; the method has high accuracy but it consumes considerable computing resources.

The vast majority of malicious document detection methods engage supervised learning, which relies on a large number of labeled samples. However, in real-world detection scenarios, it is difficult to accumulate enough malicious samples for classifier training. Additionally, the lack of suitable real samples leads to data differentiation that produces large gaps between training and testing results.

The proposed malicious PDF document detection method (3SPDF) accepts the reality that only a small number of malicious (labeled) samples are present in a large number of unknown (unlabeled) samples. The 3SPDF method extracts features from PDF document structures and divides a document file into blocks to compute entropy sequences. The wavelet energy spectrum is used to convert entropy sequences of different lengths to fixed length features. The 3SPDF method leverages semi-supervised learning in which malicious documents are treated as positive samples and small portions of unlabeled samples are treated as negative samples. To enhance the generalization capability, a random sub-sampling strategy is employed to train a set of sub-classifiers that are superimposed to detect malicious PDF documents. Experimental results demonstrate that the 3SPDF method provides 94% accuracy despite using training data with just 11% labeled malicious samples.

2. Background and Related Work

This section discusses the PDF document structure and document file entropy that underlie the proposed detection method, along with related work on malicious PDF document detection.

2.1 PDF Document Structure

Figure 1 shows the physical structure of a PDF document stored on a computer [1]. The physical structure comprises a header, body, cross-reference table (Xref) and trailer:

- **Header:** The header, which is the first line of a PDF document, includes the version number of the document. The header format is `%PDF-[Version Number]` where, for example, `%PDF-1.7` denotes a PDF 1.7 version document.

- **Body:** The document body corresponding to the main document content comprises multiple indirect objects. Indirect objects store text, pictures and other content, including fonts, colors, text sizes, etc. An indirect object starts with its object number, generation number and keyword `obj` and ends with the keyword `endobj`. A character stream starts with the keyword `stream` and ends with the keyword `endstream`. The contents of the dictionary are stored as a list of key-value pairs (attributes and attribute values). Character streams generally store text content, picture information and embedded JavaScript code. Compression methods are employed to reduce PDF document storage space and transmission costs.

- **Cross-Reference Table:** The cross-reference table starts with the keyword `xref`. This section mainly records the offset addresses of objects in the document relative to the starting position. The unit of storage is a byte.

- **Trailer:** The trailer starts with the keyword `trailer` and ends with the string `%%EOF`. The trailer stores two important pieces of information. One is the root object identified by the attribute keyword `Root` and the other is the main role. The byte offset position of the cross-reference table relative to the starting point of the document is identified by the keyword `startxref`.

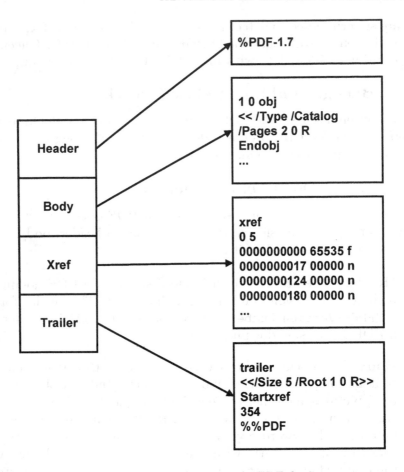

Figure 1. Physical structure of a PDF document.

2.2   Document Entropy

Entropy is widely used in document analysis [3]. In this work, it expresses uncertainty within bytes. Specifically, it sums the entropy values of the frequencies of occurrence of byte values (0x00-0xFF) observed in a data block of a given length. The entropy $H(X)$ of a data block X is computed as:

$$H(X) = -\sum_{i=1}^{255} p(i) \times \log_2 p(i)$$

where $p(i)$ is the probability that a byte value i appears in data block X.

Editing a document using a popular tool such as Microsoft Word or Adobe Acrobat causes parameters such as control symbols in the document to fluctuate within the range of the entropy values. When a document has been modified to trigger malicious behavior, the PDF structure is usually abnormal and this is manifested in its entropy values. For example, a stack overflow is triggered by adding a series of meaningless placeholder characters in a PDF document, which causes the entropy values to be outside the normal range.

2.3   Malicious PDF Document Detection

Malicious PDF document detection typically relies on signatures or heuristic rules [14]. Hash values computed for content fragments are compared against a fingerprint library to identify anomalies. This approach is labor intensive and is difficult to scale.

Machine learning techniques are increasingly used to perform large-scale document analysis and anomaly detection. Tzermias et al. [21] have proposed a detection method that employs document content and keywords as features, but the method has shortcomings with regard to selecting and processing metadata. Gibert et al. [7] have proposed a combined static-dynamic detection method that extracts JavaScript code embedded in a PDF document and executes the code in a simulation environment to extract possible shellcode. Although this method has the advantages of dynamic detection, it is hindered by the difficulty involved in locating and extracting JavaScript code. Torres and De Los Santos [20] have developed a framework for detecting malicious PDF documents in cloud environments. Zhang [27] has proposed a multilayer perceptron neural network model with stochastic gradient descent search for model updates. The model, which relies on supervised learning, requires large numbers of labeled positive and negative samples. Unfortunately, as mentioned above, in real-world malicious document detection scenarios, it is difficult to accumulate enough malicious document samples for classifier training.

3.   Malicious PDF Document Detection Method

Figure 2 provides an overview of the proposed 3SPDF method for detecting malicious PDF documents. The method has two phases, feature extraction and classification.

In the initial feature extraction phase, each document sample is parsed based on its keywords to produce a structure feature vector. The document sample is also divided into blocks by a sliding window to compute entropy sequences. The wavelet energy spectrum is used to convert the

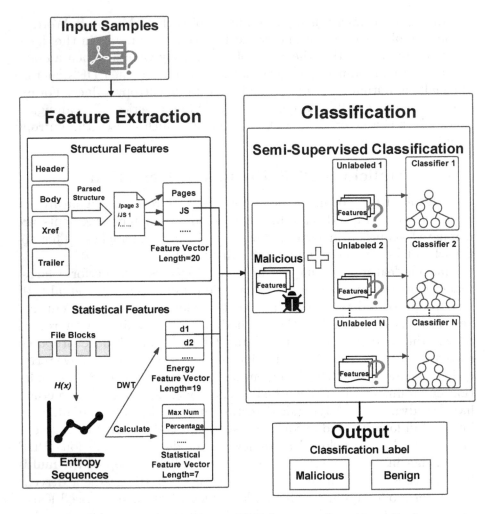

Figure 2. 3SPDF malicious PDF document detection method.

varying sequence lengths to fixed length features. Additionally, global statistical features are computed based on the entropy sequences.

In the subsequent classification phase, the structural and statistical features of malicious and unlabeled document samples are used to train several base classifiers. The collection of trained base classifiers is then used to analyze test samples and output the detection results.

3.1   Structural Features

Figure 3 shows a malicious PDF document sample. The code on the left-hand side executes a malicious `cmd.exe` file using the launch action.

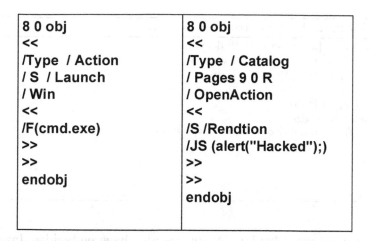

Figure 3. Malicious PDF document.

An attacker can use JavaScript code on the right-hand side to execute malicious actions.

The information in the sample PDF document is different from that in a normal PDF document because malicious code is embedded in the document. Therefore, the following structural features are employed in malicious PDF document detection:

- **/Page:** This feature indicates the number of pages in a PDF document. It is used as an auxiliary feature because many malicious PDF documents have just one page, but some benign PDF documents also have one page; additionally, malicious PDF documents may have multiple pages.

- **/Encrypt:** This feature indicates if a PDF document requires a password to be read or has a digital certificate that controls document rights.

- **/ObjStm:** This feature indicates the number of object streams in a PDF document. Some object streams contain objects that obfuscate malicious code. Also, objects (including URLs) may be present in an object stream.

- **/JS and /JavaScript:** These features indicate that a PDF document contains embedded JavaScript code. Almost all malicious PDF documents have embedded JavaScript code that performs actions such as downloading malware or executing heap overflows.

Algorithm 1: Extract structural features.

Input: D (= $\{d_i\}$): Set of PDF documents,
K (= $\{k_i\}$): Set of structural keywords
Output: F (= $\{f_{i,j}\}$): Set of features
$F = \emptyset$
for *each d_i in D* **do**
    for *k_j in K* **do**
        $f_{i,j} \leftarrow$ number of occurrences of k_j in d_i
        $F = F \cup \{f_{i,j}\}$
    end
end
return F

However, some benign documents also have embedded JavaScript code.

- **/AA and /OpenAction:** These features indicate that automated actions are executed when a PDF document is opened using a browser or document reader. Malicious attackers often execute such actions in combination with JavaScript code. When a victim opens the malicious PDF document, the JavaScript code runs automatically via OpenAction without the victim's knowledge.

  The combination of JavaScript code and automatic actions is suspicious. Many malicious PDF documents use this combination to execute embedded JavaScript code or download malicious programs from remote servers.

Algorithm 1 is employed to extract the structural features from PDF documents.

3.2   Entropy-Based Statistical Features

The entropy of a PDF document is computed by dividing it into blocks using a fixed-size sliding window of 256 bytes. If the size of the last block is not larger than 128 bytes, then the block is dropped from consideration. If the size of the last block is between 128 and 256 bytes, then it is padded with zeroes.

Assume that a PDF document D has N blocks: $\{X_0, X_1, \ldots, X_{N-1}\}$. Then, the entropy $H(X_k)$ of each block X_k is computed as:

$$H(X_k) = -\sum_{i=1}^{255} p(i) \times \log_2 p(i)$$

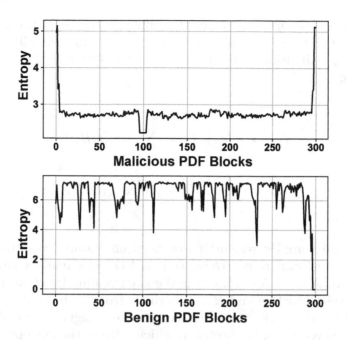

Figure 4. Entropy sequences for malicious and benign PDF documents.

where $p(i)$ is the probability of a byte value i appearing in data block X_k. The entropy values range from 0 through 8. The entropy value is 0 when all the bytes are the same and the entropy value is 8 when all the bytes are different.

Figure 4 shows the entropy sequences for malicious and benign PDF documents. Note that a malicious document has different entropy sequences from a benign document. However, the sizes of the PDF documents are different. In the example, the longest entropy sequence could be 1,551 whereas the shortest entropy sequence may be just 52. Different feature lengths render machine learning ineffective. Therefore, the wavelet energy spectrum is used to change the varying sequence lengths to fixed length features.

The discrete wavelet transform is commonly used in signal and image processing applications. In this work, entropy sequences are regarded as signals over time. Unlike the Fourier transform, the discrete wavelet transform can capture information about location and frequency that are important characteristics of entropy sequences.

During wavelet decomposition, an entropy sequence is decomposed into approximation and detail coefficients. The approximation coefficients represent the overall contour of a signal and retain the main information of the signal. The detail coefficients represent portions of the signal that change drastically along with other signal details.

The approximation coeficients $a_{j,k}$ and detail coefficients $d_{j,k}$ are computed as:

$$a_{j,k} = \langle \phi_{j,k}, D \rangle = \sum_{t=0}^{N-1} \phi_{j,k}(t) H(X_t)$$

$$d_{j,k} = \langle \psi_{j,k}, D \rangle = \sum_{t=0}^{N-1} \psi_{j,k}(t) H(X_t)$$

where j and k are the resolution and position parameters of the discrete wavelet transform, respectively, D is a PDF document containing N blocks $\{X_0, X_1, \ldots, X_{N-1}\}$, $\phi_{j,k}$ is the Haar scaling function, $\psi_{j,k}$ is the Haar wavelet function and $H(X_t)$ is the entropy of block X_t in D.

The discrete wavelet transform decomposes a signal into multiple levels. The wavelet energy spectrum, which reflects the entropy value distribution, is chosen to express the features associated with entropy sequences. In particular, the energy E_j is computed from the detail coefficients $d_{j,k}$ as:

$$E_j = \sum_{k=1}^{2^j - 1} (d_{j,k})^2$$

where j and k are the resolution and position parameters of the discrete wavelet transform, respectively.

The maximum level of the discrete wavelet transform depends on the size of the PDF document. A sample with N points in the entropy sequence can have $J = \log_2 N$ features extracted in the wavelet energy spectrum. This work employed 19 features, which implies that the maximum document size is no more than $2^{19} \times 256$ bytes (about 16 MB). This was not an issue because the PDF document samples in the dataset were less than 16 MB in size. A PDF document with less than 19 features was padded with zeroes. Also, as mentioned above, variable entropy length sequences were converted to fixed length features.

Statistical features for malicious PDF document detection were also extracted from the original entropy sequences. Table 1 describes the extracted statistical features. The features help analyze entropy sequences in a holistic manner, which enhances the ability to detect malicious PDF documents.

Table 1. Statistical features.

Feature	Description
Length	Length of entropy sequence
MSE	Mean squared error (MSE) of entropy sequence
Max	Maximumm value of entropy sequence
Zero-Percentage	Percentage of entropy sequences with zero values
Max-Percentage	Percentage of entropy sequences with values greater than 7.0

3.3   Classification

After the PDF document features are extracted, semi-supervised machine learning is employed to train a classifier. Since the training data comprises a small number of malicious (positive) samples in a large population of unlabeled samples, a classifier is created using the positive-unlabeled bagging techniques [15]. Specifically, all the malicious samples and random unlabeled samples are employed for training. Out-of-bag estimates are recorded during training. The process is repeated a number of number of times and each sample is assigned the average of its out-of-bag estimates.

Multiple independent decision tree sub-classifiers are constructed to increase the generalization capability. The sub-classifiers are superimposed by computing the averages of their individual scores during malicious PDF document detection. This approach is more robust to noise than one employing a support vector machine.

Algorithm 2 specifies the positive-unlabeled bagging procedure. It employs a bagging strategy to construct multiple sub-classifiers using the training dataset. The average of the sub-classifier probability scores is computed to label test samples as positive (1) or benign (0). The algorithm is very effective when the training set has a limited number of positive (malicious) samples.

4.   Experiments and Results.

Three experiments were performed to evaluate the efficacy of the proposed 3SPDF method at detecting malicious PDF documents. The first experiment evaluated the performance of the feature set. The second experiment analyzed 3SPDF performance against two other positive-unlabeled classifiers. The third experiment compared the performance of the 3SPDF method against three popular malicious PDF detection methods.

Algorithm 2: Positive-unlabeled bagging.

Input: M $(= \{m_i\}$ $i = 0, \ldots$, K-1): Set of K malicious (positive) samples,
U $(= \{u_j\}$ $j = 0, \ldots$, L-1 (L $>$K)): Set of unlabeled samples,
N: number of iterations
Output: S $(=\{s(x_j)\}$ where $s(x_j) = f(x_j)/n(x_j)$ for x_j in U)
for $j=0$ to L-1 **do**
    $n(u_j)=0$, $f(u_j)=0$
end
for $t=1$ to N **do**
    Draw a sample U_t of size K from U
    Train decision tree classifier f_t using positive samples M and negative
    samples U_t
    f_t returns 1 if the test sample is positive else it returns 0
    $U' = U - U_t$ (i.e., samples in U not used in training)
    for *each element x_j in U'* **do**
        if *f_t applied to x_j returns positive* **then** $f(x_j) \leftarrow f(x_j) + 1$
        $n(x_j) \leftarrow n(x_j) + 1$
    end
end
for $j=0$ to L-1 **do**
    $s(x_j) = f(x_j)/n(x_j)$
end
return S

4.1   Dataset Creation and Experimental Setup

A total of 19,815 PDF documents were collected from VirusTotal [23], VirusShare [22] and Contaigo [5]. Several researchers have created datasets from the resources maintained at these websites.

The dataset created in this research comprised 10,815 known malicious PDF documents and 9,000 known benign PDF documents with sizes ranging from 2 KB to 5 MB. Malicious PDF documents tested by VirusTotal included malicious JavaScript code and/or open actions and other malicious behaviors. The benign PDF documents included articles, official documents, reports, etc.

The computer used in the experiments was a Windows 10 Professional 64 bit machine with an Intel i7-8550U processor, 8 GB RAM and 256 GB SSD. The training dataset included labeled malicious PDF documents and unlabeled (malicious and benign) PDF documents.

After extracting the features, each PDF document was described as a 46-dimensional vector. Min-max normalization was employed to convert the features to the same scale. A decision-tree-based method was employed to train the positive-unlabeled learning models.

Table 2. Feature set performance.

Features	Precision	Recall	F1-Score	Accuracy
Structural	90.62%	90.67%	90.70%	90.66%
Statistical	92.50%	91.55%	91.55%	91.51%
Combined	94.77%	94.75%	94.75%	94.74%

4.2 Evaluation Metrics

Malicious samples were designated as positive whereas benign samples were designated as negative. The true positive TP metric considers malicious samples classified as positive samples. The true negative TN metric considers malicious samples classified as negative samples. The false positive FP metric considers benign samples classified as positive samples. The false negative FN metric considers benign samples classified as negative samples.

The overall Precision, Recall, Accuracy and F1-Score metrics were computed as follows:

$$\text{Precision} = \frac{TP}{TP + FP}$$

$$\text{Recall} = \frac{TP}{TP + FN}$$

$$\text{Accuracy} = \frac{TP + TN}{TP + TN + FN + FP}$$

$$\text{F1-Score} = 2 \times \frac{\text{Precision} \times \text{Recall}}{\text{Precision} + \text{Recall}}$$

4.3 Feature Set Analysis

Two types of features were employed for classifier training and detection: structural features and statistical features. The first experiment was conducted to evaluate feature set performance. The training set comprised 6,500 samples: 1,000 labeled malicious samples and 5,500 unlabeled (malicious/benign) samples.

Table 2 shows the feature set performance results. Using only structural features to train the 3SPDF classifier yielded consistent performance slightly above 90% for all four metrics whereas using only statistical features to train the 3SPDF classifier yielded performance between 91.51% and 92.50% for all four metrics. In contrast, combining the structural and statistical features yielded the best results – between 94.74% and 94.77% for all four metrics.

Attackers often create malicious PDF documents with similar structural features as benign PDF documents. For example, they may add extra pages to malicious documents so that the documents have similar numbers of pages as benign documents or they may create malicious documents by modifying benign documents. The statistical features based on entropy sequences highlight artificial changes to documents. The results in Table 2 demonstrate that combining statistical features with structural features when training the 3SPDF classifier enhances detection performance.

4.4 Classifier Analysis

The second experiment analyzed 3SPDF classifier performance against the performance of two positive-unlabeled classifiers. One was a standard classifier that used unlabeled samples as negative samples. The other classifier employed a two-step approach [8].

The datasets used in the experiment were created by randomly dividing the original dataset into training and testing datasets with 6,500 and 13,315 samples respectively. The training dataset had 3,250 malicious samples and 3,250 benign samples.

To assess the positive-unlabeled learning performance, all 3,250 benign samples in the training dataset were considered to be unlabeled samples. An adjustable parameter `hidden_size` was used to specify the number of malicious samples in the training dataset that were considered to be unlabeled samples. For example, a `hidden_size` of 500 implies that the training dataset had 2,750 positive samples and 3,750 unlabeled samples (with 500 malicious and 3,250 benign samples). The larger the `hidden_size` value, the fewer the number of positive samples in the training dataset.

In the experiment, datasets of different scales were used to train the classifiers. The `hidden_size` parameter value was varied from 500 to 3,000 in steps of 300. A decision tree model was used for all the base classifiers with the number of estimators set to 100.

Figure 5 compares classifier performance for increasing `hidden_size` values. At the beginning of the experiment (low values), all three classifiers exhibit good detection performance. This is because the vast majority of unlabeled samples were benign samples and the malicious samples in the unlabeled samples did not impact performance. Indeed, the learning process is close to conventional supervised learning.

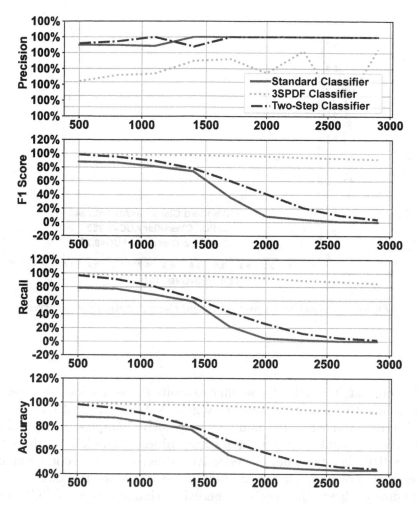

Figure 5. Classifier performance for increasing `hidden_size` values.

As the `hidden_size` values increase, more malicious samples are unlabeled and the proportions of positive samples in the training sets decrease. The precision values for all three classifiers remain at high levels. However, the other three performance metrics drop for the standard and two-step classifiers. In fact, when the `hidden_size` value is around 3,000, the recall and F1-score fall sharply to 0% – this implies that the two classifiers are less capable of detecting malicious samples. Note that when the `hidden_size` values are in the 500 to 2,000 range, the standard and two-step classifiers only detected small numbers of malicious samples that were obviously different from the benign samples. This is why the two classifiers have low recall but high precision scores.

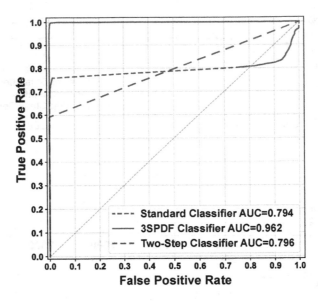

Figure 6.   Classifier receiver operating characteristic (ROC) curves.

In contrast, the 3SPDF classifier exhibits stable performance regardless of how the training dataset changes. Notably, the 3SPDF classifier is effective even when the numbers of positive samples in the training datasets are small (large `hidden_size` values). The key result is that the 3SPDF classifier would be effective in real-world detection scenarios where few labeled malicious samples are available for training.

Figure 6 shows the receiver operating characteristic (ROC) curves for the three classifiers with the `hidden_size` parameter set to 1,800. The 3SPDF classifier performs the best and is most stable. Also, the area under curve (AUC) values for the standard and two-step classifiers are 0.794 and 0.796, respectively, whereas the 3SPDF classifier has a significantly higher value of 0.962. Indeed, the 3SPDF classifier takes full advantage of the structural and statistical features, and yields excellent results.

The 3SPDF classifier was also evaluated using unbalanced training datasets that are common in real-world scenarios. The unbalanced training datasets in the experiment were created by varying the percentages of malicious samples from 11.5% to 34.6%. The results in Table 3 demonstrate that the 3SPDF classifier exhibits excellent (93.98% to 98.77%) scores for all four performance metrics even when provided with unbalanced training datasets.

Table 3. Unbalanced dataset performance.

Malicious Samples	Precision	Recall	F1-Score	Accuracy
11.5%	94.68%	93.98%	94.00%	93.98%
19.2%	95.95%	95.57%	95.59%	95.57%
34.6%	98.77%	98.75%	98.75%	98.74%

4.5 Detection Method Comparison

The third experiment was conducted to compare the performance of the 3SPDF method against three popular malicious PDF detection methods, Isolation Forest, PDFrate [17] and Hidost [19].

Table 4. Comparison of detection method performance.

Detection Method	Precision	Recall	F1-Score	Accuracy
Isolation Forest	55.41%	53.66%	40.77%	55.46%
PDFrate	95.34%	95.94%	95.63%	95.37%
Hidost	97.72%	99.67%	98.94%	97.75%
3SPDF (value=0)	97.55%	99.60%	98.56%	98.74%
3SPDF (value=1)	99.69%	98.10%	98.89%	98.74%
3SPDF (average)	98.77%	98.75%	98.75%	98.74%

Table 4 compares the effectiveness of the four methods for malicious PDF document detection. All the scores obtained by the 3SPDF method are above 97%, which are comparable or better than the scores obtained by the PDFrate and Hidost methods that use supervised learning. In contrast, Isolation Forest, which uses unsupervised learning, exhibits poor performance (precision of 55.41%) due to the absence of adequate labeled data.

The PDFrate and Hidost methods only consider structural features. This is problematic because malicious PDF documents are often created to look like benign documents and, therefore, may evade detection based on structural features. The 3SPDF method exhibits good classification performance compared with the PDFrate and Hidost methods despite being trained with a small number of malicious samples; 3SPDF's use of semi-supervised learning reduces its dependence on correctly-labeled training samples. Indeed, the 3SPDF method also has the best performance because it leverages structural and statistical features in malicious PDF document detection.

Table 5.   Comparison of detection method time requirements.

Detection Method	Time
Isolation Forest	3.84 s
PDFrate	7.21 s
Hidost	5.23 s
3SPDF	3.87 s

It is important to note that the accuracy of the 3SPDF method can be improved further. This is because a detailed examination of the dataset samples revealed that some malicious PDF documents contained malicious URLs, which were not reflected in the features employed by the 3SPDF method.

Table 5 shows the time required by each of the four methods to detect 10,000 malicious PDF document samples. The 3SPDF method result is better than, albeit relatively close to, the results obtained with the other methods that use supervised/unsupervised learning. A key advantage is that the 3SPDF method relies on static features (unlike dynamic detection methods) so the time requirements are low, rendering it very attractive in real-time malicious PDF document detection applications.

5.   Conclusions

The 3SPDF method described in this chapter effectively leverages semi-supervised machine learning to detect malicious PDF documents. Traditional supervised-learning-based detection methods that rely heavily on labeled samples of malicious documents are relatively ineffective in real-world scenarios due to the limited availability of labeled malicious document samples. In contrast, the semi-supervised learning model employed by the 3SPDF method leverages labeled as well as unlabeled samples to effectively classify malicious and benign PDF documents. Experimental results demonstrate that 3SPDF yields an accuracy of 94% using training data containing just 11% of labeled malicious samples.

A key advantage of the 3SPDF method arises from its use of structural and statistical features in malicious PDF document detection. This is important because many malicious PDF documents are created to have similar structural features as benign PDF documents. Statistical features based on entropy sequences highlight artificial changes made to documents, so combining statistical features with structural features significantly enhances detection performance. Another advantage is the use of static feature learning, which unlike dynamic detection methods, has low time requirements; this renders the 3SPDF method attractive in real-time malicious PDF document detection applications.

Finally, a promising feature of the 3SPDF method is that, with simple modifications, it can be used to detect other types of malicious documents, including Microsoft Office documents.

Acknowledgement

This research was supported by the National Natural Science Foundation of China under Grant no. 61572469.

References

[1] Adobe Systems, Document Management – Portable Document Format – Part 1: PDF 1.7, First Edition 2008-7-1, PDF 32000-1:2008, First Edition 2008-7-1, San Jose, California, 2008.

[2] A. Blonce, E. Filiol and L. Frayssignes, Portable Document Format (PDF) security analysis and malware threats, presented at the *Black Hat Europe Conference*, 2008.

[3] G. Canfora, F. Mercaldo and C. Visaggio, An HMM and structural entropy based detector for Android malware: An empirical study, *Computers and Security*, vol. 61, pp. 1–18, 2016.

[4] A. Cohen, N. Nissim, L. Rokach and Y. Elovici, SFEM: Structural feature extraction methodology for the detection of malicious office documents using machine learning methods, *Expert Systems with Applications*, vol. 63, pp 324–343, 2016.

[5] Contaigo, 16,800 Clean and 11,960 Malicious Files for Signature Testing and Research (contagiodump.blogspot.com/2013/03/16800-clean-and-11960-malicious-files.html), March 24, 2013.

[6] FireEye, Advanced Persistent Threat Groups, Milipitas, California (www.fireeye.com/current-threats/apt-groups.html), 2020.

[7] D. Gibert, C. Mateu, J. Planes and R. Vicens, Classification of malware by using structural entropy on convolutional neural networks, *Proceedings of the Thirty-Second AAAI Conference on Artificial Intelligence, Thirtieth AAAI Conference on Innovative Applications of Artificial Intelligence and Eighth AAAI Symposium on Educational Advances in Artificial Intelligence*, pp. 7759–7764, 2018.

[8] A. Kaboutari, J. Bagherzadeh and F. Kheradmand, An evaluation of two-step techniques for positive-unlabeled learning in text classification, *International Journal of Computer Applications Technology and Research*, vol. 3(9), pp. 592–594, 2014.

[9] M. Li, Y. Liu, M. Yu, G. Li, Y. Wang and C. Liu, FEPDF: A robust feature extractor for malicious PDF detection, *Proceedings of the IEEE International Conference on Trust, Security and Privacy in Computing and Communications*, pp. 218–224, 2017.

[10] J. Lin and H. Pao, Multi-view malicious document detection, *Proceedings of the Conference on Technologies and Applications of Artificial Intelligence*, pp. 170–175, 2013.

[11] L. Liu, X. He, L. Liu, L. Qing, Y. Fang and J. Liu, Capturing the symptoms of malicious code in electronic documents by file entropy signals combined with machine learning, *Applied Soft Computing*, vol. 82, article no. 105598, 2019.

[12] X. Lu, F. Wang and Z. Shu, Malicious Word document detection based on multi-view feature learning, *Proceedings of the Twenty-Eighth International Conference on Computer Communications and Networks*, 2019.

[13] D. Maiorca, D. Ariu, I. Corona and G. Giacinto, A structural and content-based approach for precise and robust detection of malicious PDF files, *Proceedings of the International Conference on Information Systems Security and Privacy*, pp. 27–36, 2015.

[14] D. Maiorca, G. Giacinto and I. Corona, A pattern recognition system for malicious PDF file detection, *Proceedings of the Eighth International Conference on Machine Learning and Data Mining in Pattern Recognition*, pp. 510–524, 2012.

[15] F. Mordelet and J. Vert, A bagging SVM to learn from positive and unlabeled examples, *Pattern Recognition Letters*, vol. 37, pp. 201–209, 2014.

[16] J. Muller, F. Ising, V. Mladenov, C. Mainka, S. Schinzel and J. Schwenk, Practical decryption exfiltration: Breaking PDF encryption, *Proceedings of the ACM SIGSAC Conference on Computer and Communications Security*, pp. 15–29, 2019.

[17] C. Smutz and A. Stavrou, Malicious PDF detection using metadata and structural features, *Proceedings of the Twenty-Eighth Annual Computer Security Applications Conference*, pp. 239–248, 2012.

[18] N. Srndic and P. Laskov, Detection of malicious PDF files based on hierarchical document structure, *Proceedings of the Twentieth Annual Network and Distributed System Security Symposium*, 2013.

[19] N. Srndic and P. Laskov, Hidost: A static machine-learning-based detector of malicious files, *EURASIP Journal on Information Security*, vol. 2016(1), article no. 45, 2016.

[20] J. Torres and S. De Los Santos, Malicious PDF document detection using machine learning techniques, *Proceedings of the Fourth International Conference on Information Systems Security and Privacy*, pp. 337–344, 2018.

[21] Z. Tzermias, G. Sykiotakis, M. Polychronakis and E. Markatos, Combining static and dynamic analysis for the detection of malicious documents, *Proceedings of the Fourth European Workshop on System Security*, article no. 4, 2011.

[22] VirusShare, Home (`www.virusshare.com`), 2020.

[23] VirusTotal, GUI (`www.virustotal.com/gui`), 2020.

[24] M. Xu and T. Kim, PlatPal: Detecting malicious documents with platform diversity, *Proceedings of the Twenty-Sixth USENIX Security Symposium*, pp. 271–287, 2017.

[25] W. Xu, Y. Qi and D. Evans, Automatically evading classifiers: A case study on PDF malware classifiers, *Proceedings of the Twenty-Third Annual Network and Distributed Systems Security Symposium*, 2016.

[26] M. Yu, J. Jiang, G. Li, C. Lou, Y. Liu, C. Liu and W. Huang, Malicious document detection for business process management based on a multi-layer abstract model, *Future Generation Computer Systems*, vol. 99, pp. 517–526, 2019.

[27] J. Zhang, MLPdf: An effective machine learning based approach for PDF malware detection, presented at *Black Hat USA*, 2018.

Chapter 8

MALICIOUS LOGIN DETECTION USING LONG SHORT-TERM MEMORY WITH AN ATTENTION MECHANISM

Yanna Wu, Fucheng Liu and Yu Wen

Abstract   Advanced persistent threats routinely leverage lateral movements in networks to cause harm. In fact, lateral movements account for more than 80% of the time involved in attacks. Attackers typically use stolen credentials to make lateral movements. However, current detection methods are too coarse grained to detect lateral movements effectively because they focus on malicious users and hosts instead of abnormal log entries that indicate malicious logins.

This chapter proposes a malicious login detection method that focuses on attacks that steal credentials. The fine-grained method employs a temporal neural network embedding to learn host jumping representations. The learned host vectors and initialized attribute vectors in log entries are input to a long short-term memory with an attention mechanism for login feature extraction, which determines if logins are malicious. Experimental results demonstrate that the proposed method outperforms several baseline detection models.

Keywords: Malicious login detection, LSTM with attention mechanism

1.   Introduction

Advanced persistent threats (APTs) have been the focus of considerable research [4, 8]. These threats manifest themselves as sustained and effective attacks against specific targets. A perpetrator typically compromises a host and adopts a hiding strategy to enter a sleep state. After the presence on the host is consolidated, information is collected about the targeted network. Lateral movements, the next crucial and time-consuming phase of the attack, involve attempts to progressively move to and control other machines in the network [24]. Advanced per-

© IFIP International Federation for Information Processing 2021
Published by Springer Nature Switzerland AG 2021
G. Peterson and S. Shenoi (Eds.): Advances in Digital Forensics XVII, IFIP AICT 612, pp. 157–173, 2021.
https://doi.org/10.1007/978-3-030-88381-2_8

sistent threat actors almost always perform lateral movements by steal-
ing credentials [20]. Consequently, malicious login detection is vital to
combating advanced persistent threats.

Some approaches for detecting lateral movements model logins in communication graphs that describe interactions between users and hosts [1,
4, 13]. However, these approaches primarily focus on partial properties
of log entries (e.g., login relationships) to identify malicious users and
hosts, which is a relatively coarse-grained detection strategy.

Other lateral movement detection techniques leverage machine learning to obtain better results [2, 18]. However, most techniques are limited by model interpretability. Additionally, some fine-grained detection
methods (e.g., [5, 23]) do not consider interactions between hosts. These
issues make it difficult to develop effective machine-learning-based classifiers that can detect lateral movements.

This chapter proposes a malicious login detection method that focuses
on attacks that steal credentials. The method involves host representation learning and feature extraction from log files. Host representation
learning engages a temporal neural network embedding model to learn
the initial host vectors, enabling the translation of host login relationships to host vectors. The host vectors and initial expressions of other
attributes are the inputs for log feature extraction. The proposed feature
extraction model learns log entry vectors using long short-term memory
(LSTM) coupled with an additional attention mechanism that enhances
the extraction of information about important attributes. The log vectors are subsequently input to a multilayer perceptron (neural network
with a full connection layer) that classifies them as malicious or benign.

The proposed malicious login detection method incorporates some
novel features. Inspired by research in text classification, long short-
term memory is employed to learn information about attributes and
extract the meaning of a log file. Instead of merely detecting malicious
hosts and users, each log entry is analyzed and classified as malicious
or not, allowing for fine-grained detection. The attention mechanism
emphasizes important attributes and strengthens the interpretability of
the model. Additionally, since malicious logins occur between hosts,
host interactions are considered using a temporal graph embedding to
learn preferred host representations and integrate them in the log vectors. Experimental results demonstrate that the malicious login detection method has a false positive rate of just 0.002% and outperforms
several state-of-the-art detection models.

2.   Related Work

In recent years, considerable research has focused on advanced persistent threat detection, especially during the lateral movement phase. Siadati et al. [21] developed APT-Hunter, a visualization tool for exploring login data to discover patterns and detect malicious logins. They advocated the use of machine learning models to enhance feature extraction. Bai et al. [2] evaluated a number of supervised and unsupervised machine learning methods for detecting lateral movements, including Logistic Regression, Gaussian Naive Bayes, Decision Tree, Random Forest, LogitBoost, feed-forward neural networks and clustering methods. However, their results were modest because insufficient testing data was available.

Chen et al. [7] developed a network embedding approach that uses an autoencoder to detect lateral movements. However, the performance was limited by data imbalance – the available red team activity data was much less than normal activity data. Holt et al. [9] employed deep autoencoders to detect lateral movements. Three autoencoder models were developed that addressed the data imbalance problem to some extent, but the results were limited by the paucity of abnormal data. The three models achieved an average recall in excess of 92% and an average false positive rate less than 0.5%. Bohara et al. [4] focused on broader patterns of system interactions after attackers gain initial access to hosts. Their graph-based target system model yielded true positive and false positive rates of 88.7% and 14.1%, respectively.

Malicious logins are key components of cyber attacks. As a result, much research has focused on malicious login detection, especially in internal networks. Some detection approaches analyze user activities. Powell [16] modeled the normal daily behavior of users in behavior graphs; alerts were raised when deviations in user behavior were detected. Kent et al. [13] employed bipartite authentication graphs to analyze user behavior in enterprise networks. Amrouche et al. [1] applied knowledge discovery techniques to detect malicious login events. Other detection methods focus on host activities. For example, Chen et al. [7] analyzed host communications whereas Bian et al. [3] examined host authentication logs.

Regardless of whether the methods mentioned above focused on user or host activities, they detected anomalies using partial attributes in login information. To address this limitation, researchers have analyzed login events using rule-based [20] and machine-learning-based approaches [5, 23]. However, malicious login detection has been hindered by the lack of model interpretability. Additionally, the methods do not

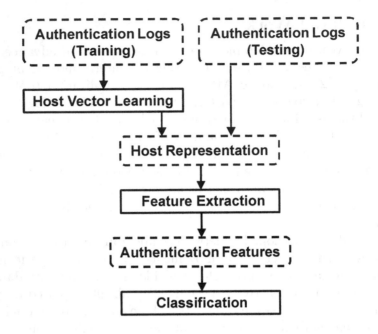

Figure 1. Proposed malicious login detection method.

focus on host interactions. In contrast, the proposed fine-grained malicious login detection method employs temporal neural network embedding to learn host jumping representations. The learned host vector and other initialized attribute vectors in log entries are then fed to a long short-term memory with an attention mechanism for login feature extraction, which helps determine whether or not logins are malicious.

3.   Preliminaries

This section provides an overview of the malicious login detection method and its threat model.

3.1   Detection Method Overview

The proposed malicious login detection method focuses on jumps between hosts and login information. Inspired by the success of neural networks in text classification [6, 15, 25], the proposed method trains a classifier to distinguish between benign and malicious authentication log activities.

Figure 1 provides an overview of the proposed detection method. The method has three main phases, host vector learning, feature extraction and classification. Host vector learning employs the Hawkes temporal network embedding (HTNE) model [26] to learn host characteristics in space and time from a host connection graph. Feature extraction employs a long short-term memory model with an attention mechanism to capture the semantics of logs. Specifically, a login entry is expressed as: "a source user employs a destination user's certificate to log into a target host from the source host." Drawing on research in text classification, log attributes are considered to be text that is input to a feature extraction process to produce feature vectors. The final classification phase employs a multilayer perceptron (fully-connected network) to classify login operations as benign or malicious.

3.2 Threat Model

The proposed method focuses on credential-based lateral movements. In this paradigm, attackers achieve lateral movements by establishing footholds with stolen login credentials.

The threat model assumes that login data can be collected in a timely manner and that the login data is not tampered with during transmission. Also, new login data is continually introduced to enable the model to re-learn the host initial vectors.

4. Proposed Method

This section describes the host vector learning approach, feature extraction model, attention mechanism and classification model optimization.

4.1 Host Vector Learning

A lateral movement involves a jump from one host to another. Lateral movements by a user are modeled in a jump graph in which nodes are hosts and jumps are edges. A jump between hosts is a jump in space and time. This is a typical problem in temporal network embedding learning, which means that the representation vectors of hosts can be learned.

The proposed method uses the Hawkes temporal network embedding (HTNE) model [26] to create a temporal graph embedding node representation. The method learns the interactions between hosts and explores the time sequences of host interactions. It integrates the Hawkes process in network embedding to capture the impacts of previous neighbor nodes on the current neighbor node.

The login dataset used in this research records time in seconds. At this granularity, the amount of data is huge. Therefore, the times were

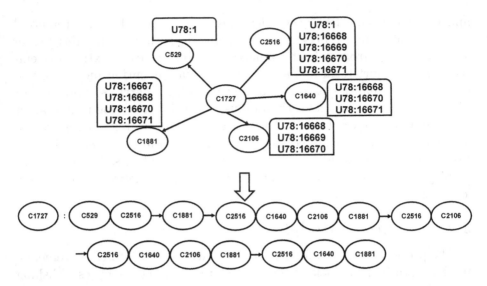

Figure 2.   Host temporal network and neighborhood formation sequence.

converted to minutes and the temporal network was constructed in the order of minutes.

Figure 2 presents an example of a host temporal network and neighborhood formation sequence. The top portion of the figure shows a temporal network whose nodes are hosts. In addition to interactions between hosts, interaction time sequences exist for source users.

The bottom portion of Figure 2 shows neighborhood formation sequences initiated by source hosts. Each neighborhood formation sequence is a chronological sequence of the same user. After a sequence is created, it is modeled using the multi-dimensional Hawkes process and the model is trained as in [26] to obtain the vector corresponding to each node, specifically, the representation of each host.

4.2   Feature Extraction

Unlike previous work on malicious login detection that has focused on the macro-level entities (i.e., users or hosts), this research considers micro-level entities (i.e., log entries). This fine-grained approach requires the pure host-based representation to be modified. Each log entry has the form node-edge-node corresponding to (source host – login – target host). Specifically, the "login" information from the log entries has to be embedded in the edges.

Unfortunately, it is difficult to learn the representations of nodes and the expressions of edges in a temporal network graph. Additionally, handling the large number of edge combinations requires substantial computing resources. Therefore, a neural network model is used to acquire the login information.

A neural network offers three advantages when attempting to learn login information. First, a neural network can handle a large number of log entries generated in a complex computer network environment. Second, lateral movements have subtle (non-obvious) rules and features; a neural network can automatically extract features and adapt to a dynamic computer network environment. Third, malicious login detection methods are plagued by large numbers of false positives (i.e., benign logs are classified as malicious); the powerful feature extraction ability of a neural network enables the false positive rate to be reduced.

Combined with the characteristics of the dataset itself, the processed authentication log is essentially in a language with a grammar and each log entry is a sentence that conforms to the grammar. In fact, each log entry is translated to: "the source user logs into the target host using the credentials of the target user from the source host via some authentication type and login type." This is similar to the text classification problem in the field of natural language processing.

Research has shown that using a recurrent neural network for the feature extraction task is significantly better than using a convolutional neural network. However, a recurrent neural network has gradient disappearance and gradient explosion problems. Therefore, a long short-term memory neural network is employed by the proposed method. Figure 3 shows the long short-term memory based feature extraction model.

At this point, the feature extraction phase can be specified. Let $X = [x_1, x_2, ..., x_7]$ denote a log entry, where x_1 corresponds to source user, x_2 to destination user, x_3 to source host, x_4 to destination host, x_5 to authentication type, x_6 to login type and x_7 to authentication orientation (login/logoff).

The interactive host representation produced by the host vector learning phase and the other attribute representations are randomly initialized into a long short-term memory neural network for extracting the login characteristics. Next, each log attribute x_i is input to a two-layer long short-term memory neural network and the hidden vector h_i is produced for each attribute as follows:.

$$[h_1, h_2, ..., h_7] = \text{LSTM}([x_1, x_2, ..., x_7])$$

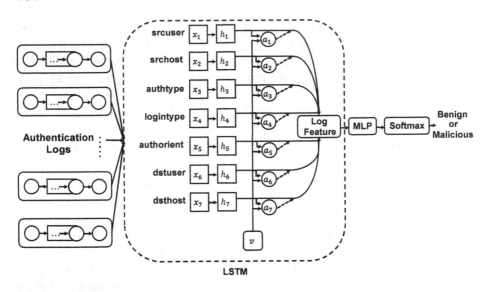

Figure 3. Feature extraction model.

4.3   Attention Mechanism

Not all attributes contribute equally to the meaning of a log entry. Therefore, an attention mechanism is incorporated to extract the attributes that are important to the meaning of a log entry and aggregate the representation of the informative attributes to create a log vector.

First, the hidden vector of each log attribute h_i obtained via feature extraction is passed to the activation function tanh to yield the attention score (importance) u_i of the log attribute:

$$u_i = v^T tanh(W_h \cdot h_i + W_b)$$

where W_h and W_b are the weight and bias parameters, respectively, and v is an artificially-defined parameter vector. These parameters are learned using the stochastic gradient descent technique.

The normalized attention score α_i of the log attribute is computed as:

$$\alpha_i = \frac{exp(u_i)}{\sum_{j=1}^{n} exp(u_j)}$$

where n is the number of attributes in a log.

Next, the log entry vector log_i corresponding to the log attribute is computed as:

$$log_i = \sum_{j=1}^{n} \alpha_j \cdot h_j$$

The log entry vector log_i is passed to a multilayer perceptron (MLP) with a full connection layer for supervised learning and classification, and the result is input to the softmax function to compute the probability p_{log_i} of the log entry:

$$p_{log_i} = \text{softmax}(\text{MLP}(log_i))$$

4.4   Classification Model Optimization

After obtaining the probability of a log entry p_{log_i}, the cross-entropy loss function is selected as the target function and used for training:

$$loss = -\sum_{i=1}^{N} log_i^{label} \cdot \log p_{log_i}$$

where log_i^{label} is the true label (benign or malicious) of the log entry, $\log p_{log_i}$ is the logarithm of the log entry probability p_{log_i} and N is the number of log entries.

Note that the *label* of each log entry in the training dataset is assigned to be benign or malicious during the training phase.

During the testing phase, two probability values corresponding to benign and malicious are obtained after processing by the multilayer perceptron and softmax function. Each log entry is classified as benign or malicious depending on which of the two has the higher probability value.

5.   Experimental Evaluation

This section describes the experiments conducted to evaluate the performance of the proposed malicious login detection method along with the evaluation results.

5.1   Dataset Description

The experiments drew on the Comprehensive, Multi-Source Cyber-Security Events Dataset from Los Alamos National Laboratory [12]. The dataset comprises five data elements with units of seconds: (i) Windows-based authentication events, (ii) process start and stop events, (iii) DNS lookups, (iv) network flow data and (v) well-defined red team events. The dataset, which is approximately 12 GB in compressed form, contains more than 1.6 billion events associated with 12,425 users and 17,684

Table 1. Dataset statistics.

Dataset	Total Logs	Malicious Logs
Training Dataset	3,67,6507	340
Testing Dataset	6,213,591	409

computers. The content includes authentication logs over 58 consecutive days. Although the data was collected in a real environment, some data is missing due to equipment errors or other causes [17].

In order to use valid data and avoid flawed data, only the authentication events and red team activities were employed for training and testing. The red team events, which corresponded to typical advanced persistent threat activity, were labeled as malicious. The red team events comprised 749 malicious logins involving 98 compromised users.

All the malicious logins involving the 98 compromised users occurred during the month of January. Therefore, the dataset used in the experiments included the logs for the entire month of January. Logs from January 1 through January 12 were used for training and the logs from January 13 through January 31 were used for testing.

Table 1 shows the compositions of the training and testing datasets. The 749 malicious logs in the training and testing datasets came from the red team file. Analysis of the datasets revealed that 203 of the 409 malicious logins in the testing dataset were not in the training dataset. The logs used in the experiments corresponded to <source user, source host, target host> triples.

Each authentication event in the datasets comprised nine attributes: time, source user, destination user, source computer, destination computer, authentication type, login type, authentication orientation and success/failure. The login success/failure attribute was eliminated because failed logins were ignored. Also, the time attribute was excluded during feature extraction to obtain a pure categorical feature space.

5.2   Experimental Setup

The computing system used in the experiments was an Nvidia Tesla V100 GPU wth three cores, each with 16 GB display memory. The model was constructed using `pytorch1.4.0` and the anomalous login detection method was programmed in `python3.6.8`.

The pre-processing vectors had attributes with 64 dimensions. During the training phase, it was determined that two-layer stacked LSTMs performed better than a one-layer LSTM and were not as good as three-layer stacked LSTMs. However, two-layer stacked LSTMs were selected to save time and adjust the model quickly after adding new user data. The hyper-parameters used for training were dropout = 0.5, batch size = 128 and hidden size = 64. Preliminary experiments revealed that the learning rate was better when its initial value was set to one.

5.3 Evaluated Models

The malicious login detection method was compared against several baseline models. To demonstrate the utility of the attention mechanism two versions of the proposed method were employed. The proposed method, incorporating long short-term memory with an attention mechanism, is referred to as MLDLA. The second version, employing long short-term memory without the attention mechanism, is referred to as MLDL.

The following five baseline models were evaluated in the experiments:

- **Tiresias [19]:** Tiresias leverages recurrent neural networks to predict future events based on previous observations. It learns user login actions and predicts the next event using long short-term memory.

- **Ensemble [4]:** The Ensemble model relies on a graph of hosts and uses an ensemble of two anomaly detectors to identify compromised hosts. It employs three steps to detect attacks that rely on lateral movements: feature extraction, feature analysis and anomaly detection. The model uses network flow logs, and command and control and lateral movement traces to create a communication graph. The model is coarse grained, but it is a good unsupervised learning representative.

- **Bagging Machine Learning [11]:** Bagging ML is a supervised learning model designed to extract advanced composite features. It leverages three models, Random Forest, LogitBoost and Logistic Regression, in a majority voting configuration. The model uses authentication events and red team activities. It was applied to 21 users selected from 98 malicious users. The training dataset comprised the first 12.5 days of data totaling 199,090 events and the testing dataset comprised 25,900 events containing 37 batches.

- **Semi-Supervised Outlier Detection [10]:** The SSOD model employs an automatic semi-supervised outlier ensemble detector whose automatic feature is supported by unsupervised outlier ensemble theory. It has two phases, training dataset preparation and semi-supervised ensemble construction. It employed a random

Table 2.   Model performance.

Method	TPR	FPR	Accuracy	AUC	F1-Score
Tiresias	99.75%	66.24%	33.76%	5.3%	0.2%
Ensemble	91.15%	13.95%	86.05%	89.17%	–
Bagging ML	100%	0.19%	99.62%	–	65.87%
SSOD	100%	0.17%	–	–	–
Log2vec	100%	10%	–	91%	–
MLDL	98.28%	0.0024%	99.99%	99.99%	74.17%
MLDLA	98.78%	0.002%	99.99%	99.99%	76.08%

sample of 150,000 consecutive authentication events containing at least five malicious events in its evaluation.

- **Log2vec [14]:** Log2vec is a heterogeneous graph embedding based modularized method that proposes several rules to construct a graph from which the log character representation can be learned. It uses clustering to distinguish between malicious and benign log entries Log2vec considers process events, authentication events and red team activities. Fifty users from among the 98 malicious users were selected to construct the rule-based graph.

5.4   Evaluation Results

Table 2 shows the model performance results. Note that the accuracy of the Bagging ML approach is the balanced average accuracy reported in [11]. Although the true positive rate (TPR) of the proposed MLDLA method does not reach 100%, the method used all the malicious logs, which was more reliable than the other models that only used subsets of the malicious logs. The MLDLA method has the best false positive rate (FPR) of 0.002%. Specifically, only 127 out of 6,213,182 benign logs were misclassified as malicious. The MLDLA method also has the best F1-score of 76.08%. Comparison of the MLDLA and MLDL method results demonstrates the efficacy of the attention mechanism.

Figure 4 shows the true positive and false positive rates for the MLDLA (dark lines) and MLDL (light lines) methods. The true positive and false positive rates were computed for each training epoch. The MLDLA method outperforms the MLDL method for both metrics. Taking the averages of the ten epochs with the best performance, the MLDLA true positive rate is greater than 98% whereas the MLDL rate is 96%. In the best case, MLDLA misclassified 127 logs as malicious whereas MLDL misclassified 140 logs as malicious.

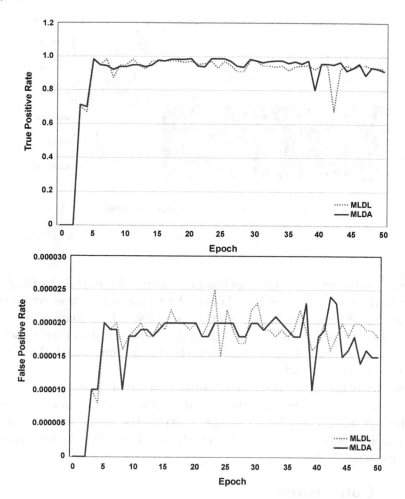

Figure 4. True positive and false positive rates for the MLDLA and MLDL methods.

In addition to improving detection performance, the attention mechanism enhances the interpretability of the model. Figure 5 shows six logs (rows), three benign and three malicious. The shades of the cells express the relative weights of the log attributes (darker shades correspond to greater weights). Clearly, the malicious and benign logs have different weights.

5.5 Optimization and Learning Rate

The stochastic gradient descent and Adam optimizers were used in the experiments. The best results were obtained with the stochastic gradient

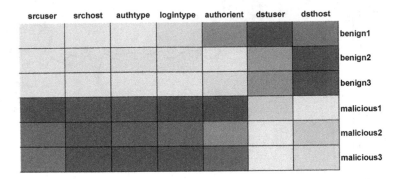

Figure 5.   Attribute attention in logs.

descent optimizer. For this optimizer, setting the initial learning rate to one achieved better results. When the learning rate was set to 0.01, all the logs in the testing dataset were classified as benign (i.e., 409 malicious logs were classified as benign). Although it is rare to set the initial learning rate to one, the experiment demonstrated that it was effective for the dataset.

The Adam optimizer essentially employs the RMSprop algorithm with a momentum term. Setting the initial learning rate to one resulted in all the logs being classified as benign. Setting the initial learning rate to 0.001 yielded a false positive rate of 0.0014%. However, the true positive rate was only 76%.

6.   Conclusions

Advanced persistent threats routinely leverage lateral movements in networks to cause harm. Lateral movements typically involve malicious logins using stolen credentials. However, current detection methods that focus on malicious users and/or hosts are too coarse grained to detect lateral movements effectively. In contrast, the fine-grained method presented in this chapter focuses on abnormal log entries. It employs a temporal neural network embedding to learn host jumping representations. The learned host vector and initialized attribute vectors in log entries are input to a long short-term memory with an attention mechanism for login feature extraction, which help determine if logins are malicious. Experimental results demonstrate that the proposed method has a true positive rate in excess of 98% and a false positive rate of 0.002%. It also outperforms several baseline detection models, especially with regard to the false positive rate and F1-score.

Future research will evaluate the performance of the proposed method on other datasets. Additionally, it will attempt to engage unsupervised learning to account for the paucity of labeled malicious data.

References

[1] F. Amrouche, S. Lagraa, G. Kaiafas and R. State, Graph-based malicious login events investigation, *Proceedings of the IFIP/IEEE Symposium on Integrated Network and Service Management*, pp. 63–66, 2019.

[2] T. Bai, H. Bian, A. Daya, M. Salahuddin, N. Limam and R. Boutaba, A machine learning approach for RDP-based lateral movement detection, *Proceedings of the Forty-Fourth IEEE Conference on Local Computer Networks*, pp. 242–245, 2019.

[3] H. Bian, T. Bai, M. Salahuddin, N. Limam, A. Daya and R. Boutaba, Host in danger? Detecting network intrusions from authentication logs, *Proceedings of the Fifteenth International Conference on Network and Service Management*, 2019.

[4] A. Bohara, M. Noureddine, A. Fawaz and W. Sanders, An unsupervised multi-detector approach for identifying malicious lateral movement, *Proceedings of the Thirty-Sixth IEEE Symposium on Reliable Distributed Systems*, pp. 224–233, 2017.

[5] A. Brown, A. Tuor, B. Hutchinson and N. Nichols, Recurrent Neural Network Attention Mechanisms for Interpretable System Log Anomaly Detection, arXiv: 1803.04967 (arxiv.org/abs/1803.04967), 2018.

[6] H. Chen, M. Sun, C. Tu, Y. Lin and Z. Liu, Neural sentiment classification with user and product attention, *Proceedings of the Conference on Empirical Methods in Natural Language Processing*, pp. 1650–1659, 2016.

[7] M. Chen, Y. Yao, J. Liu, B. Jiang, L. Su and Z. Lu, A novel approach for identifying lateral movement attacks based on network embedding, *Proceedings of the IEEE International Conference on Parallel and Distributed Processing with Applications, Ubiquitous Computing and Communications, Big Data and Cloud Computing, Social Computing and Networking, and Sustainable Computing and Communications*, pp. 708–715, 2018.

[8] I. Ghafir, M. Hammoudeh, V. Prenosil, L. Han, R. Hegarty, K. Rabie and F. Aparicio-Navarro, Detection of advanced persistent threat using machine learning correlation analysis, *Future Generation Computer Systems*, vol. 89, pp. 349–359, 2018.

[9] R. Holt, S. Aubrey, A. DeVille, W. Haight, T. Gary and Q. Wang, Deep autoencoder neural networks for detecting lateral movement in computer networks, *Proceedings of the International Conference on Artificial Intelligence*, pp. 277–283, 2019.

[10] G. Kaiafas, C. Hammerschmidt, S. Lagraa and R. State, Auto semi-supervised outlier detection for malicious authentication events, in *Machine Learning and Knowledge Discovery in Databases*, P. Cellier and K. Driessens (Eds.), Springer, Cham, Switzerland, pp. 176–190, 2020.

[11] G. Kaiafas, G. Varisteas, S. Lagraa, R. State, C. Nguyen, T. Ries and M. Ourdane, Detecting malicious authentication events trustfully, *Proceedings of the IEEE/IFIP Network Operations and Management Symposium*, 2018.

[12] A. Kent, Comprehensive, Multi-Source Cyber-Security Events, Los Alamos National Laboratory, Los Alamos, New Mexico (`csr.lanl.gov/data/cyber1`), 2015.

[13] A. Kent, L. Liebrock and J. Neil, Authentication graphs: Analyzing user behavior within an enterprise network, *Computers and Security*, vol. 48, pp. 150–166, 2015.

[14] F. Liu, Y. Wen, D. Zhang, X. Jiang, X. Xing and D. Meng, Log2vec: A heterogeneous graph embedding based approach for detecting cyber threats within an enterprise, *Proceedings of the ACM SIGSAC Conference on Computer and Communications Security*, pp. 1777–1794, 2019.

[15] P. Liu, X. Qiu and X. Huang, Recurrent Neural Network for Text Classification with Multi-Task Learning, arXiv: 1605.05101 (`arxiv.org/abs/1605.05101`), 2016.

[16] B. Powell, Detecting malicious logins as graph anomalies, *Journal of Information Security and Applications*, vol. 54, article no. 102557, 2019.

[17] M. Pritom, C. Li, B. Chu and X. Niu, A study on log analysis approaches using the Sandia dataset, *Proceedings of the Twenty-Sixth International Conference on Computer Communications and Networks*, 2017.

[18] T. Schindler, Anomaly Detection in Log Data Using Graph Databases and Machine Learning to Defend Advanced Persistent Threats, arXiv: 1802.00259 (`arxiv.org/abs/1802.00259`), 2018.

[19] Y. Shen, E. Mariconti, P. Vervier and G. Stringhini, Tiresias: Predicting security events through deep learning, *Proceedings of the ACM SIGSAC Conference on Computer and Communications Security*, pp. 592–605, 2018.

[20] H. Siadati and N. Memon, Detecting structurally-anomalous logins within enterprise networks, *Proceedings of the ACM SIGSAC Conference on Computer and Communications Security*, pp. 1273–1284, 2017.

[21] H. Siadati, B. Saket and N. Memon, Detecting malicious logins in enterprise networks using visualization, *Proceedings of the IEEE Symposium on Visualization for Cyber Security*, 2016.

[22] G. Tang, M. Muller, A. Rios and R. Sennrich, Why Self-Attention? A Targeted Evaluation of Neural Machine Translation Architectures, arXiv: 1808.08946 (`arxiv.org/abs/1808.08946`), 2018.

[23] A. Tuor, R. Baerwolf, N. Knowles, B. Hutchinson, N. Nichols and R. Jasper, Recurrent Neural Network Language Models for Open Vocabulary Event-Level Cyber Anomaly Detection, arXiv: 1712.00557 (`arxiv.org/abs/1712.00557`), 201.

[24] L. Yang, P. Li, Y. Zhang, X. Yang, Y. Xiang and W. Zhou, Effective repair strategy against advanced persistent threat: A differential game approach, *IEEE Transactions on Information Forensics and Security*, vol. 14(7), pp. 1713–1728, 2019.

[25] Z. Yang, D. Yang, C. Dyer, X. He and E. Hovy, Hierarchical attention networks for document classification, *Proceedings of the Conference of the North American Chapter of the Association for Computational Linguistics: Human Language Technologies*, pp. 1480–1489, 2017.

[26] Y. Zuo, G. Liu, H. Lin, J. Guo, X. Hu and J. Wu, Embedding temporal network via neighborhood formation, *Proceedings of the Twenty-Fourth ACM SIGKDD International Conference on Knowledge Discovery and Data Mining*, pp. 2857–2866, 2018.

IV

NOVEL APPLICATIONS

Chapter 9

PREDICTING THE LOCATIONS OF UNREST USING SOCIAL MEDIA

Shengzhi Qin, Qiaokun Wen and Kam-Pui Chow

Abstract   The public often relies on social media to discuss and organize activities such as rallies and demonstrations. Monitoring and analyzing open-source social media platforms can provide insights into the locations and scales of rallies and demonstrations, and help ensure that they are peaceful and orderly.

This chapter describes a dictionary-based, semi-supervised learning methodology for obtaining location information from Chinese web forums. The methodology trains a named entity recognition model using a small amount of labeled data and employs n-grams and association rule mining to validate the results. The validated data becomes the new training dataset; this step is performed iteratively to train the named entity recognition model. Experimental results demonstrate that the iteratively-trained model has much better performance than other models described in the research literature.

Keywords: Social media analysis, location extraction, named entity recognition

1.   Introduction

Since 2019, large-scale protests by the Anti-Extradition Law Amendment Bill Movement (Anti-ELAB Movement) have occurred in Hong Kong [13]. Predicting the locations and scales of such protests can assist law enforcement in planning and mobilizing resources to ensure that the protests are peaceful and orderly.

Social media is often used to organize public events [16]. Online discussion sites, including web forums, have played key roles in organizing flash mobs, protest campaigns and demonstrations during periods of unrest [1]. The monitoring and analysis of public opinion in online discussions is an effective means for obtaining information that could assist public safety efforts.

© IFIP International Federation for Information Processing 2021
Published by Springer Nature Switzerland AG 2021
G. Peterson and S. Shenoi (Eds.): Advances in Digital Forensics XVII, IFIP AICT 612, pp. 177–191, 2021.
https://doi.org/10.1007/978-3-030-88381-2_9

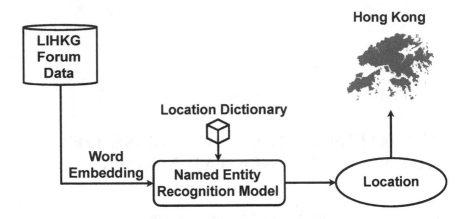

Figure 1. Location extraction methodology.

When discussing political activities in web forums, organizers often mention the locations of future rallies, such as the "818 Victoria Park Rally" [4]. Web forum users also share location information in their discussions. Therefore, it would be effective to focus on locations in web forum discussions to gain advance information about possible rallies.

This research focuses on the LIHKG public web forum [10], one of the most active platforms for discussing the Anti-ELAB Movement [8]. A dictionary-based, semi-supervised learning methodology was developed to automatically extract location information from harvested web forum data.

Figure 1 presents an overview of the location extraction methodology. The first step is to crawl a web forum to gather data. Next, the topic and post contents are processed by a named entity recognition model to identify location data. Note that the location dictionary is used in conjunction with the named entity recognition model to improve performance. Finally, the extracted locations are analyzed with other information to predict the locations where unrest may occur.

Empirical experiments have revealed that accurate location identification is the principal challenge. Another challenge is that the focus is on Chinese web forums. Unlike in English text, locations are difficult to extract because Chinese has neither word segmentation nor capitalization rules. Additionally, web forum data differs from data in structured sources such as newspapers and papers. Specifically, the language is irregular, often does not follow grammar rules and incorporates large amounts of slang and abbreviations. The dictionary-based, semi-supervised learning method developed in this research enhances the extraction of location information from Chinese web forums.

2.   Related Work

Analysis of public opinion can provide law enforcement with alerts before incidents occur as well as important information about the incidents. Tang and Song [14] proposed a visual analysis methodology for high-frequency words and user responses on the Weibo social media platform in Mainland China. Yang and Ng [15] extracted and clustered key information from social networks to classify public opinion and enhance subsequent analysis. People exchange information about the times and locations of events using social media. Because time has a standard format, it can be extracted using simple methods such as regular expressions. However, location extraction is more difficult because it does not have a standard structure.

Machine learning and deep learning methods are used to identify entities, including locations, in sentences [9]. The tagging methodologies employ hidden Markov models [3], maximum entropy Markov models [12] and conditional random fields [7]. Hammerton [5] used a long short-term memory (LSTM) based neural network model for entity recognition. Collobert et al. [2] combined deep learning with data mining methods to achieve high accuracy. Huang et al. [6] proposed one of the best-performing English entity recognition models that uses a bidirectional long short-term memory and conditional random fields; the model is robust and does not have any special dependence on word embedding.

However, considerable differences exist between English and Chinese, especially with regard to sentence structure. In English, words are separated by spaces and proper nouns such as locations and names are capitalized; Chinese does not have such features. Another problem is that named entity recognition models focus on processing structured text such as reports and news, and are not applicable to short messages in web forums. Additionally, many expressions in web forums do not follow grammar rules. Furthermore, many posts contain non-standard expressions, including abbreviations and slang terms. These characteristics render manual labeling of training data and model training much more difficult.

Zhang and Yang [17] proposed the Lattice-LSTM model, which adds word segmentation results prior to word information and fuses word information into character information; experiments conducted with the Weibo platform yielded a 62.56% match ratio. Liu [11] designed an encoding strategy based on word-character long short-term memory that improves on the Lattice-LSTM model, but it cannot be batched due to its uneven structure. Zhu et al. [18] proposed a multi-task long short-term memory method for Chinese named entity recognition. It uses

Table 1.   LIHKG web forum topic and post.

Topic:	2019-08-19 13:00:00,94, 神仙,56,反對修訂引渡條例大遊行集中討論	
Post:	2019-08-18 15:02:00,108,241242, 明天旺角集合	
Field	**Value**	**English Translation**
Topic Created Time	2019-08-19 13:00:00	
Topic Author ID	94	
Topic Author Name	神仙	Immortal
Topic ID	56	
Topic Title	反對修訂引渡條例大遊行	Anti-ELAB Protest
Post Created Time	2019-08-18 15:02:00	
Post Author ID	108	
Post Author Name	241242	
Post Content	明天旺角集合	Tomorrow Mong Kok meet up

three long short-term memory structures to model related information –
one is based on character information for named entity recognition, the
second is based on word information for word segmentation tasks and
the third is based on a shared structure that models character and word
information. The method achieved 59.31% accuracy on Weibo.

All the methods presented above require the manually labeling of large
amounts of data. It is difficult to manually label forum data because web
forum text has many spoken expressions and does not follow grammar
rules, resulting in different people labeling entities differently. These
differences require an additional verification step. In contrast, the proposed training method using dictionary-based, semi-supervised learning
requires only a small amount of data to be labeled manually and uses
prior location information to extract locations in web forum posts.

3.   Location Extraction from Web Forum Data

This section describes the dictionary-based, semi-supervised learning
methodology for extracting locations from web forum data.

3.1   Web Forum Dataset

This work has focused on LIHKG, a popular web forum website that
is often referred to as the Hong Kong version of Reddit. LIHK was one
of the most active platforms for discussing the Anti-ELAB Movement.
Chinese language data was collected from the Current Affairs channel in
LIHKG, the principal channel for discussing political events and protest
movements. Table 1 shows a sample LIHKG web forum post.

Table 2. Locations in dictionary and web forum posts.

Dictionary Content	Web Forum Content
旺角 (Mong Kok)	旺角港鐵站 (Mong Kok MTR Station)
維多利亞公園 (Victoria Park)	維園 (Vi-Park)
皇后大道中 (Queen's Road Central)	皇后大道中168號 (Queen's Road Central #168)

The dataset used in this research comprised more than 300,000 LIHKG forum data items collected from 302,109 posts between August 18, 2019 to October 10, 2019. The posts, which had 42,258 distinct authors, contained 2,126 topics.

3.2 Dictionary-Based Semi-Supervised Learning

In real-world applications, dictionary-based keyword matching is often used to identify locations because it is simple, convenient and fast, and has low false positive rates. However, it cannot recognize locations outside the dictionary especially when the locations mentioned in web forums are not specified completely. Table 2 shows examples of locations used in web forum posts that do not completely match locations in the dictionary.

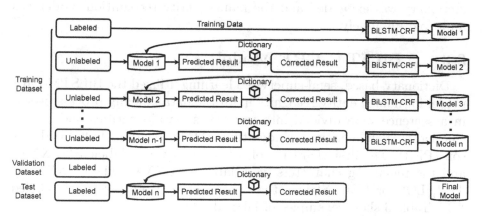

Figure 2. Dictionary-based, semi-supervised learning method.

Figure 2 shows the proposed dictionary-based, semi-supervised learning method, which combines named entity recognition and dictionary matching. The dataset was divided into three parts: a training dataset, a validation set and a testing dataset.

The training dataset was divided into several parts, named $part_1$, $part_2$, $part_3$, etc. Each part contained 50,000 posts. Only data in $part_1$ was labeled (i.e., only 50,000 posts were labeled). The validation set contained 2,000 posts for intermediate corrections. The testing dataset contained another 2,000 posts for final testing.

The labeled training data ($part_1$) was input to a bidirectional long short-term memory and conditional random field (BiLSTM-CRF) model for training. Model training was performed iteratively. Note that $model_1$ denotes the first trained model and $model_i$ denotes the model trained after the ith iteration. Each iteration involved three steps:

- **Step 1:** Apply $model_i$ to predict the 50,000 posts in $part_{i+1}$. The dictionary is applied in conjunction with the n-Gram-ARM algorithm (described below) to correct the training results.

- **Step 2:** Label the new corrected results using BIO labeling and re-input them as training data to $model_i$ to obtain $model_{i+1}$.

- **Step 3:** Compare all the identified locations with the current dictionary. If a location is not in the dictionary, then it is added to the dictionary as a new location.

The iterations were repeated until all the parts of the training dataset were utilized. By continuously updating the model and dictionary, the dictionary was expanded and the named entity recognition model was improved iteratively.

3.3   BiLSTM-CRF Model

Dictionary-based, semi-supervised learning utilized the BiLSTM-CRF model for named entity recognition. In the data labeling process, words in a sentence were divided into two types – valid entities and invalid characters. Valid entities included time (TIM), location (LOC) and person (PER). The first character of a valid entity was labeled "B-XXX" and the remaining characters were labeled "I-XXX" where "XXX" is TIM, LOC or PER. Strings that were not valid entities were labeled "O." Table 3 shows examples of labeled text.

Upon being given a Chinese post, the BiLSTM-CRF model automatically identified the beginning and end of a location. Figure 3 shows the BiLSTM-CRF model structure. In addition to identifying location information in the post, the time and person entities were labeled in order to learn the relationships between entities and tags.

Word embedding was performed to convert text into a vector for input to the BiLSTM-CRF model. LSTM (long short-term memory) is

Table 3.   Labeled text example.

English Translation	Tomorrow		Mong	Kok	meet	up
Chinese Post	明	天	旺	角	集	合
Labels	B-TIM	I-TIM	B-LOC	I-LOC	O	O

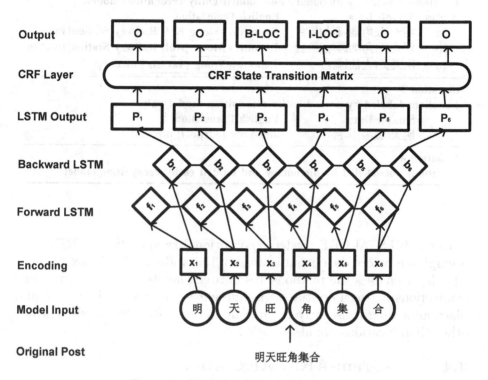

Figure 3.   BiLSTM-CRF model structure.

a deep learning model that remembers information from the previous step. However, the LSTM model can only predict the output of the next layer based on information from the previous layer. In some problems, the output of the current layer is related to the previous state and also depends on the future state. For example, a missing word in a current sentence is predicted based on the previous paragraph as well the content that comes after the current sentence. To accomplish this, the BiLSTM (bidirectional long short-term memory) model incorporates an additional layer that uses the end of a sentence as the starting input of the sequence. The BiLSTM model processes the results of the two layers to improve named entity recognition.

Table 4. Possible situations leading to location recognition results.

Situation 1		
Locations labeled by dictionary only		
	Sample Data	**English Translation**
	明天<u>旺角</u>集合	Tomorrow **Mong Kok** meet up

Situation 2		
Locations labeled by dictionary and named entity recognition model		
Cases	**Sample Data**	**English Translation**
2a	明天[<u>旺角地鐵站</u>]集合	Tomorrow **[Mong Kok Railway Station]** meet up
2b	明天<u>旺</u>[角地鐵站]集合	Tomorrow **Mong [Kok Railway Station]** meet up
2c	明天<u>黃</u>[大仙]集合	Tomorrow **Wong [Tai Sin]** meet up

Situation 3		
Locations labeled by named entity recognition model only		
	Sample Data	**English Translation**
	明天[機場]集合	Tomorrow **[Airport]** meet up

Situation 4	
Locations not labeled by dictionary and named entity recognition model	

In the BiLSTM-CRF model, a conditional random field (CRF) layer is employed before the output instead of the traditional softmax method. This layer enables the relationships between labels to be learned. Some connections between entities with different attributes, such as time and place appearing consecutively, may be discovered. This is why entities other than locations are also labeled.

3.4   *n*-Gram-ARM Algorithm

As mentioned above, the proposed methodology utilizes the location dictionary to correct the model prediction results during the iterative training process. The results obtained after the iterative training were found to have many false positives and false negatives. Also, there were deviations between the dictionary labeling results and the named entity recognition results.

Table 4 shows the possible situations leading to location recognition results using the dictionary and named entity recognition model. Dictionary labels are underlined. Named entity recognition model labels are enclosed in square brackets. Correct labels are presented in boldface. Case 2a covers situations where the named entity recognition model labeling results are supersets of the dictionary labeling results. Case 2b covers situations where the named entity recognition model and dic-

Table 5. Solutions for specific situations.

Situation	Solution
1	Keep only dictionary-labeled parts
2a	n-Gram-ARM
2b	n-Gram-ARM
2c	Keep only dictionary-labeled parts
3	Simple n-Gram-ARM
4	Iterative training

tionary labeling results have non-empty intersections. Case 2c covers situations where the named entity recognition model labeling results are subsets of the dictionary labeling results.

Table 5 shows the solutions employed to reduce the false negative rates for the various situations. As discussed above, Situation 4 is handled by the iterative training process.

The locations labeled using the dictionary can be assumed to be correct, but the dictionary may not contain all the known locations and the new locations. This problem was addressed by applying the n-Gram-ARM algorithm, which has an association rule mining (ARM) part and an n-Gram part.

The association rule mining part of the algorithm used two indicators, Support and Confidence, to measure the goodness of the labels. Support is the frequency of the union of the words in the web forum corpus labeled in the dictionary and by the named entity recognition model. Confidence is the probability of the union of dictionary-labeled and named-entity-recognition-model-labeled words if the dictionary-labeled words exist. Specifically:

$$\text{Support} = freq(Dic \cup NER)$$

$$\text{Confidence} = \frac{freq(Dic \cup NER)}{freq(Dic)}$$

where Dic denotes the words labeled as locations by the dictionary, NER denotes the words predicted as locations by the named entity recognition model and $freq$ is the number of occurrences.

Support measures how often a word appears whereas Confidence measures how likely the two parts of the label appear consecutively. The greater the values of two indicators, the greater the probability that the word is a location. For example, if the dictionary extraction result is

Table 6.   Association rule mining false positive example.

Sample Data	English Translation
明天[# **旺角地鐵站**]集合	Tomorrow [# **Mong Kok Railway Station**] meet up

"Mong Kok" and the named entity recognition model extraction result is "Kok Railway Station" (Situation 2b in Table 4), then the union of the two extracted results is "Mong Kok Railway Station." The two indicators are computed as:

$$Support = freq(旺角地鐵站)$$
$$Confidence = freq(旺角地鐵站)/freq(旺角)$$

If the Support and Confidence values are both greater than the association rule mining threshold, then the word is considered to be a location and the named entity recognition model label is retained. After training for the first time and using association rule mining, words that are not in the dictionary may be obtained. These words correspond to new locations identified by association rule mining.

The n-Gram method reduces the number of false positives. However, other false positives are possible. Specifically, words labeled by the two methods are incorrect and the labeled part of the named entity recognition model is a superset of the correct answers. Table 6 shows an example of a false positive obtained by association rule mining.

When the association rule mining validation result of the named entity recognition model output is lower than the association rule mining threshold, n-Gram-ARM is used to recheck the result. Specifically, association rule mining performed on sub-words of the named entity recognition model results in different lengths. The range of lengths n in the n-Gram method is:

$$len(Dic) < n < len(NER)$$

where $len(Dic)$ is the length of a word labeled by the dictionary and $len(NER)$ is the length of a word labeled by the named entity recognition model.

In the example above, association rule mining was performed after applying the n-Gram method. Table 7 shows the n-Gram results.

The phrases to be retained as locations were based on the association rule mining results. A higher n-Gram threshold was used to further filter the words. Since there is no dictionary labeling in Situation 3, the

Table 7. *n*-Gram results.

n-Gram Phrases for ARM with Sample Data: # 旺角地鐵站		
n = 5	n = 4	n = 3
# 旺角地鐵/ 旺角地鐵站	# 旺角地/ 旺角地鐵/ 角地鐵站	# 旺角/ 旺角地/ 角地鐵/ 地鐵站

confidence cannot be computed. Therefore, the solution for Situation 3 is named Simple *n*-Gram-ARM.

4.   Experiments and Results

Comparative experiments were conducted with several commonly-used Chinese named entity recognition models to determine the model with the best performance.

Table 8.   Chinese named entity recognition model results.

Model	Percentage	Recall	F1-Score	F1-Score (Location Only)
HMM	48.69%	46.03%	47.32%	19.40%
CRF	56.68%	58.08%	57.37%	22.57%
LSTM	48.56%	46.30%	47.41%	16.34%
CAN-NER	63.89%	54.79%	59.00%	25.81%
Lattice-LSTM	59.61%	59.45%	59.53%	24.69%
BiLSTM-CRF	62.87%	58.90%	60.82%	25.28%

Table 8 shows the experimental results. The BiLSTM-CRF model was the best performer on the dataset. The experiments revealed that locations were the main reason for the low accuracy. Table 8 shows that the best F1-Score with only the locations labeled is just 25.28%.

As mentioned above, the *n*-Gram-ARM algorithm was applied to improve the location extraction performance. The ARM thresholds were set to 7 for Support and 0.01 for Confidence based on experience.

Table 9 shows sample outputs obtained after validation using associated rule mining. None of the locations appeared in the dictionary. The locations in the last row of the table were filtered using the ARM threshold due to the low Support and Confidence values. The results in Table 9 also show that associated rule mining can expand the locations in the dictionary by identifying new location vocabulary with the

Table 9. Association rule mining validation outputs.

New Location	English Translation	Support	Confidence
金鐘站	Admiralty Station	63	0.078553616
葵涌警署	Kwai Chung Police Station	32	0.198757764
元朗西鐵站	Yuen Long West Rail Station	55	0.037826685
深圳羅湖	Shenzhen Luohu	35	0.224358974
維園	Vi-Park	560	—
元朗站# 罵村民	Yuen Long Station#Curse villagers	2	0.001375516

base locations. This also complies with the rules for naming and using locations.

Table 10. *n*-Gram-ARM algorithm result.

N-Gram Phrases	Support	Confidence
元朗站# 罵村民	2	0.0013755
元朗站# 罵村	2	0.0013755
朗站# 罵村民	2	0.0013755
元朗站# 罵	2	0.0013755
朗站# 罵村	2	0.0013755
站# 罵村民	2	0.0013755
元朗站	19	0.0130674
朗站# 罵	2	0.0013755
站# 罵村	2	0.0013755
# 罵村民	2	0.0013755
元朗站	**148**	**0.1017881**
朗站	19	0.0130674
站# 罵	2	0.0013755
# 罵村	2	0.0013755
罵村民	2	0.0013755

Table 10 shows the *n*-Gram-ARM algorithm result (row in bold font). The *n*-Gram thresholds were set to 20 for Support and 0.03 for Confidence based on experience. The application of the *n*-Gram method enables the recognized entities to be further decomposed and analyzed.

Table 11 shows that applying the *n*-Gram-ARM algorithm effectively reduced the number of false negatives. Additionally, the accuracy of location recognition is increased along with the overall experimental accuracy.

Table 11. Evaluation of dictionary n-Gram-ARM verification.

Model	Precision	Recall	F1-Score	F1-Score Overall
BiLSTM-CRF	17.24%	51.40%	25.28%	60.82%
BiLSTM-CRF + n-Gram-ARM	48.75%	36.44%	41.71%	78.08%

Table 12. Iterative training method evaluation.

Iteration	Precision	Recall	F1-Score	F1-Score Overall
0	17.24%	51.40%	25.28%	60.82%
1	54.39%	28.97%	37.80%	68.09%
2	60.00%	36.45%	45.35%	72.77%
3	58.28%	37.29%	45.48%	72.32%
4	56.24%	36.37%	44.17%	70.13%
5	54.84%	37.28%	44.39%	71.02%

Table 12 shows the effectiveness of the iterative training method. The method addresses problems posed by insufficient labeled data, low quality labeled data and irregular data. The results show that during the first three iterations, as the amount of data in the training dataset increased, the prediction results became increasingly accurate, until a steady state was reached. At the same time, the dictionary was also updated with more than 100 new locations. Not only is the proposed BiLSTM-CRF model more robust and accurate than the other named entity recognition models, but it also yields a comprehensive and richer dictionary of locations.

5. Conclusions

The location extraction methodology described in this chapter employs dictionary-based, semi-supervised learning to obtain location information from web forum data (specifically, from LIHKG, one of the most active platforms for discussing the Anti-ELAB Movement in Hong Kong). The extracted location information can assist law enforcement in planning and mobilizing resources to ensure that protests are peaceful and orderly. Experimental results demonstrate that the iteratively-trained model has much better performance than other models described in the research literature.

Future research will attempt to extract other information such as time and user behavior that would help identify relationships between posts and political movements. Research will also analyze web forum user behavior to discover user relationships and identify user roles in forums and events, such as organizers and active users.

References

[1] A. Breuer, The Role of Social Media in Mobilizing Political Protest: Evidence from the Tunisian Revolution, Discussion Paper No. 10/2012, German Development Institute, Bonn, Germany, 2012.

[2] R. Collobert, J. Weston, L. Bottou, M. Karlen, K. Kavukcuoglu and P. Kuksa, Natural language processing (almost) from scratch, *Journal of Machine Learning Research*, vol. 12(2011), pp. 2493–2537, 2011.

[3] S. Eddy, Hidden Markov models, *Current Opinion in Structural Biology*, vol. 6(3), pp. 361–365, 1996.

[4] Government of Hong Kong, Government Response to Public Meeting in Victoria Park, Hong Kong, China (`www.info.gov.hk/gia/general/201908/18/P2019081800818.htm`), August 18, 2019.

[5] J. Hammerton, Named entity recognition with long short-term memory, *Proceedings of the Seventh Conference on Natural Language Learning*, pp. 172–175, 2003.

[6] Z. Huang, X. Wei and K. Yu, Bidirectional LSTM-CRF Models for Sequence Tagging, arXiv: 1508.01991 (`arxiv.org/abs/1508.01991`), 2015.

[7] J. Lafferty, A. McCallum and F. Pereira, Conditional random fields: Probabilistic models for segmenting and labeling sequence data, *Proceedings of the Eighteenth International Conference on Machine Learning*, pp. 282–289, 2001.

[8] F. Lee, H. Liang, E. Cheng, G. Tang and S. Yuen, Affordances, movement dynamics and a centralized digital communication platform in a networked movement, to appear in *Information, Communication and Society*, 2021.

[9] J. Li, A. Sun, J. Han and C. Li, A survey of deep learning for named entity recognition, to appear in *IEEE Transactions on Knowledge and Data Engineering*, 2021.

[10] LIHKG, LIHKG Online Forum, Hong Kong, China (`lihkg.com`), 2021.

[11] W. Liu, T. Xu, Q. Xu, J. Song and Y. Zu, An encoding strategy based word-character LSTM for Chinese NER, *Proceedings of the Conference of the North American Chapter of the Association for Computational Linguistics: Human Language Technologies, Volume 1 (Long and Short Papers)*, pp. 2379–2389, 2019.

[12] A. McCallum, D. Freitag and F. Pereira, Maximum entropy Markov models for information extraction and segmentation, *Proceedings of the Seventeenth International Conference on Machine Learning*, pp. 591–598, 2000.

[13] M. Purbrick, A report on the 2019 Hong Kong protests, *Asian Affairs*, vol. 50(4), pp. 465–487, 2019.

[14] X. Tang and C. Song, Microblog public opinion analysis based on complex networks, *Journal of the China Society for Scientific and Technical Information*, vol. 31(11), pp. 1153–1163, 2012.

[15] C. Yang and T. Ng, Analyzing and visualizing web opinion development and social interactions with density-based clustering, *IEEE Transactions on Systems Man and Cybernetics – Part A: Systems and Humans*, vol. 41(6), pp. 1144–1155, 2011.

[16] T. Zeitzoff, How social media is changing conflict, *Journal of Conflict Resolution*, vol. 61(9), pp. 1970–1991, 2017.

[17] Y. Zhang and J. Yang, Chinese NER Using Lattice LSTM, arXiv: 1805.02023 (`arxiv.org/abs/1805.02023`), 2018.

[18] Y. Zhu, G. Wang and B. Karlsson, CAN-NER: Convolutional Attention Network for Chinese Named Entity Recognition, arXiv: 1904.02141 (`arxiv.org/abs/1904.02141`), 2020.

Chapter 10

EXTRACTING THREAT INTELLIGENCE RELATIONS USING DISTANT SUPERVISION AND NEURAL NETWORKS

Yali Luo, Shengqin Ao, Ning Luo, Changxin Su, Peian Yang and Zhengwei Jiang

Abstract   Threat intelligence is vital to implementing cyber security. The automated extraction of relations from open-source threat intelligence can greatly reduce the workload of security analysts. However, implementing this feature is hindered by the shortage of labeled training datasets, low accuracy and recall rates of automated models, and limited types of relations that can be extracted.

This chapter presents a novel relation extraction framework that employs distant supervision for data annotation and a neural network model for relation extraction. The framework is evaluated by comparing it with several state-of-the-art neural network models. The experimental results demonstrate that it effectively alleviates the data annotation challenges and outperforms the state-of-the-art neural network models.

Keywords: Threat intelligence, relation extraction, machine learning

1.   Introduction

Threat intelligence is evidence-based knowledge, including context, mechanisms, indicators, implications and actionable advice, about an existing or emerging menace or hazard to assets that can be used to inform decisions about the response to the menace or hazard [15]. Security analysts have become proficient at extracting indicators of compromise (IOCs). Indicators of compromise such as URLs, IP addresses, email addresses, domain names and hashes are easy to extract, but they are easily modified by attackers to bypass security measures. In any case,

© IFIP International Federation for Information Processing 2021
Published by Springer Nature Switzerland AG 2021
G. Peterson and S. Shenoi (Eds.): Advances in Digital Forensics XVII, IFIP AICT 612, pp. 193–211, 2021.
https://doi.org/10.1007/978-3-030-88381-2_10

indicators of compromise on their own cannot be expected to provide adequate cyber security.

Security analysts and policymakers need high-level threat intelligence to make critical decisions. High-level threat intelligence, such as tactics, techniques and procedures (TTPs), is extracted from a variety of sources and expressed in a form that enables further analysis and decision making. The sources are typically referred to as open-source intelligence (OSINT), which includes unstructured information collected from public resources such as research papers, newspapers, magazines, social networking sites, wikis, blogs, etc. [23].

The higher the level in the threat intelligence pyramid [21], the greater the difficulty of extracting information. Moreover, open-source intelligence resources are massive and complex, and require considerable manual analysis to obtain high-level threat intelligence. Some researchers have attempted to automate this process. However, as described below, these methods tend to focus on identifying cyber threat entities and ignore the relations between the entities. Additionally, threat intelligence relation extraction approaches often rely on rules and features developed by experts, making it difficult to deal with new entities and relations.

The proposed framework for threat intelligence relation extraction leverages distant supervision and neural networks. Distant supervision is a popular method for collecting and generating training datasets in the natural language processing domain [18]. The proposed framework uses distant supervision to generate a large amount of annotation data needed for machine learning relatively quickly. With adequate training data, a neural network can be created to effectively extract relations from unstructured, text-based, open-source intelligence resources.

The proposed framework is the first to produce a dataset for threat intelligence relation extraction. The framework is evaluated by comparing it against several state-of-the-art neural network models. The experimental results demonstrate that it effectively alleviates the data annotation challenges and outperforms the state-of-the-art neural network models.

2. Related Work

This section describes related work in the areas of threat intelligence datasets and threat intelligence information extraction.

2.1 Threat Intelligence Datasets

The demand for actionable threat intelligence, including datasets for extracting threat intelligence, is increasing. Mulwad et al. [20] designed a framework for extracting vulnerabilities and attack information from web text and translating them to a machine-understandable format, The datasets were drawn from 107 vulnerability description documents, but they are not open source.

McNeil et al. [16] developed a novel entity extraction and guidance algorithm that extracts valuable network security concepts. They acquired information from an online open-source website containing ten documents with seven entity types, and manually annotated the information to produce their dataset.

Jones et al. [7] specified a bootstrapping algorithm that extracts security entities and their relationships from textual information. They created a dataset using 62 document corpora from various cyber security websites, but the dataset is not publicly available.

Joshi et al. [9] extracted network-security-related link data from text documents and constructed experimental datasets via professional annotation. Their training dataset comprises 3,800 entities and 38,000 instances and their testing dataset contains 1,200 entities and 9,000 instances. However, the datasets are not publicly available.

Lal [10] has researched the extraction of security entities and concepts from unstructured text. More than 100 open-source reports were processed using screening and factual sampling methods to produce a dataset containing 60 common vulnerabilities and exposures (CVEs), 12 Microsoft announcements and 12 Adobe announcements. However, this dataset is not available to the public.

The literature survey reveals that threat intelligence datasets are rare and very few of them are publicly available. Therefore, this research has sought to develop an automated annotation method based on distant supervision that would enable security analysts to label open-source intelligence data quickly and efficiently as a precursor to creating threat intelligence datasets.

2.2 Threat Intelligence Information Extraction

Threat intelligence information extraction is a hot research topic. Liao et al. [12] have devised an automated technique that extracts indicators of compromise from security blogs and generates a machine-readable version for discovering inherent relationships in threat intelligence.

Lee et al. [11] have focused on discovering valuable security information and identifying emerging security event topics. Their system leverages modified topic graphs and topic discovery algorithms to discover information from open-source threat intelligence resources.

Mittal et al. [19] have attempted to obtain timely network security threats and vulnerabilities in an automated manner. Their system discovers, extracts and analyzes threat intelligence from Twitter feeds. A database in the WWW Resource Description Framework (RDF) format maintains the collected intelligence, and inference rules specified in the Semantic Web Rule Language (SWRL) process the data to produce network security threat and vulnerability information.

Tao et al. [24] have focused on the timely sharing of threat intelligence and responding to threat alerts. They applied an immune factor network algorithm based on a classification model to actively access and extract useful information from a large quantity of raw security information.

Gascon et al. [3] have attempted to discover potential relationships between pieces of different threat intelligence. They proposed a similarity algorithm based on attribute graphs to perform similarity correlations of text at different levels of granularity. However, their results were not very promising.

Traditional research on threat intelligence information extraction has focused on indicators of compromise. New research should focus on developing robust techniques for extracting high-level threat intelligence, such as tactics, techniques and procedures, from multiple sources and express it in a form that enables further analysis and decision making. Relation extraction leveraging deep learning theory from the natural language processing domain can significantly advance this line of research.

3.   Proposed Framework

This section describes the threat intelligence relation extraction framework developed in this research.

3.1   Overview

The proposed framework is designed to extract relations efficiently from unstructured open-source intelligence. It incorporates a distant supervision module that annotates unstructured data efficiently and a neural network module that extracts relations from unstructured threat intelligence. Figure 1 provides an overview of the framework workflow.

3.2   Problem Specification

Relation extraction is the elicitation of semantic relationships from unstructured text. Figure 2 shows an example of relation extraction.

The objective of relation extraction is to create a function $F : (c, R) \mapsto (e_1, e_2, r)$. The input data c is an unlabeled sentence to be processed

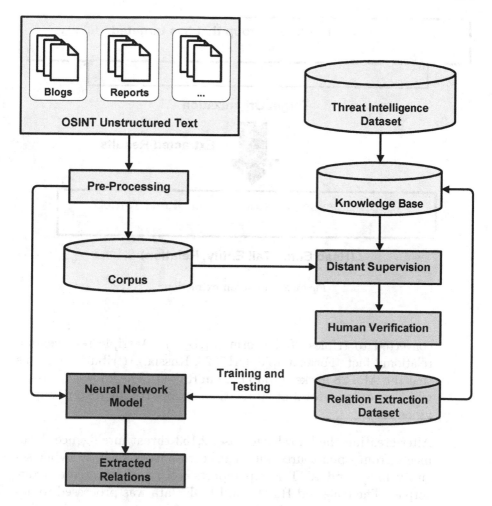

Figure 1. Framework workflow.

and $R = \{r_1, r_2, ..., r_n\}$ is a set of relations contained in sentences. The output triple, which corresponds to the prediction of c given R, comprises a head entity e_1, tail entity e_2 and relation $r \in R$ between the two entities.

3.3 Dataset

Dataset construction involves three steps: (i) knowledge base and corpus creation, (ii) distant supervision and (iii) human verification:

- **Knowledge Base and Corpus Creation:** Structured threat intelligence data conforming to the STIX II [8] specification was

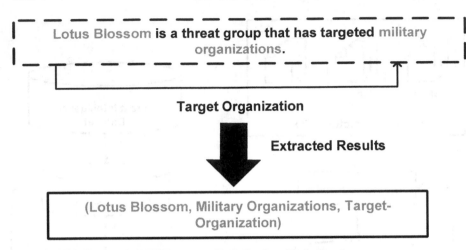

Figure 2.   Relation extraction.

converted to triples of the form (e_1, e_2, r). A triple represents a relational fact. For example, (APT28, Russia, Attribution) means that the APT28 hacker group is from Russia. The knowledge base, which contains a total of 60 relations, supports the distant supervision step.

After creating the knowledge base, 2,153 threat intelligence documents from open-source intelligence resources such as cyber security blogs and APT group reports were used to create a raw corpus. The collected HTML and PDF data was processed to obtain clean text. Named entity recognition was employed to find potential entities and co-reference resolution was used to reduce noise in the text by applying natural language processing tools such as NLTK [2], Stanford CoreNLP [14] and spaCy [22]. If a sentence contained more than one entity, then a potential relation was deemed to exist between the entities and the sentence was stored in the corpus. The final corpus contained 41,835 valid sentences.

- **Distant Supervision:** Most deep learning models require a labeled training dataset. Traditionally, a labeled training dataset is created by manually annotating training data, but this is a very time-consuming task. Another approach is to generate training data using distant supervision [18]. Distant supervision assumes

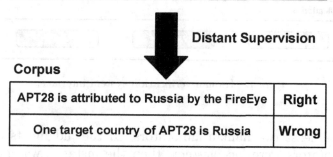

Knowledge Entities

Head Entity	Tail Entity	Relation
APT28	Russia	Attribution
...	...	...

Distant Supervision

Corpus

APT28 is attributed to Russia by the FireEye	Right
One target country of APT28 is Russia	Wrong

Figure 3. Distant supervision and noisy labeling.

that, if two entities participate in a relation, then any sentence that contains the two entities might express the relation.

The candidate set was created by considering each sentence c in the corpus. If c contained a head entity e_1 and tail entity e_2, and a triple (e_1, e_2, r) existed in the knowledge base, then it was assumed that the sentence c mentioned relation r and the tuple (e_1, e_2, r, c) was added as an instance in the candidate set. The final candidate set contained 11,906 instances.

Although distant supervision is effective at labeling data automatically, it suffers from the noisy labeling problem (shown in Figure 3). Unlike natural language processing, the cyber security domain requires strict data labeling, so all the instances in the candidate set had to be manually verified. Fortunately, the manual verification workload was reduced considerably because the candidate set was generated via distant supervision.

- **Human Verification:** Human annotators with expertise in computer science were recruited to eliminate incorrectly-labeled instances. A crowdsourcing verification platform similar to Amazon's Mechanical Turk [1] was employed.

Figure 4 shows the human verification system interface. It implements checks on the sentences and their relation triples that were labeled by distant supervision. Each instance was verified

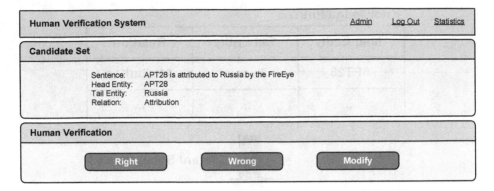

Figure 4. Human verification system interface.

by at least two human annotators. If the judgments of the two annotators were inconsistent, then the instance was passed to a third individual for proofreading and the final label determined according to the majority principle.

After the human verification, relations with less than 200 instances were eliminated to create a clean relation extraction dataset. The final dataset contained 9,277 instances, 7,035 unique entities, 18 unique relations, 8,027 entity pairs and 8,150 relational facts.

3.4   Neural Network Model

A neural network model was employed for relation extraction. The backbone of the model is a bidirectional long short-term memory (Bi-LSTM) network [4] with selective attention [13] to learn the representations of text-expressing relations. Figure 5 shows the neural network model for relation extraction. The neural network model has four layers: (i) embedding layer, (ii) encoding layer, (iii) selection layer and (iv) classification layer:

- **Embedding Layer:** For each sentence c, the Word2vec technique [17] was used to train word embeddings to project each word token onto d_w-dimensional space. The words that appeared more than 10 times in the corpus were retained as vocabulary. Position embedding [25] was also employed for all the words in each sentence to create d_p-dimensional vectors with entity position information.

- **Encoding Layer:** A Bi-LSTM neural network model was employed for each sentence encoding. Hochreiter and Schmidhu-

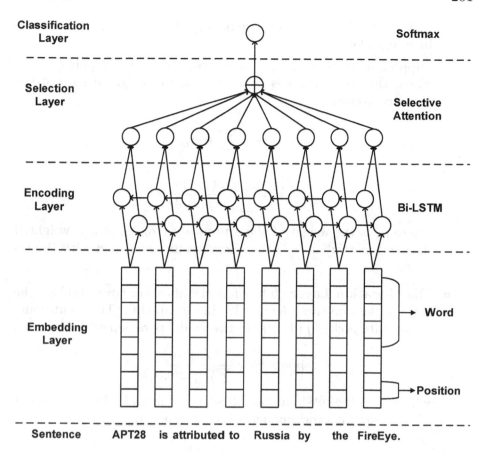

Figure 5. Neural network model for relation extraction.

ber [5] proposed the LSTM (long short-term memory) neural network model to address the gradient vanishing problem. However, a standard LSTM model only process sequences in temporal order. The Bi-LSTM neural network model [4] improves on the standard LSTM model by incorporating a second layer to obtain information from the past and future. The model is well-suited to sequence-oriented tasks such as name entity recognition and relation extraction.

■ **Selection Layer:** The selection layer employs the selective attention mechanism proposed by Lin et al. [13]. It uses sentence-level attention to select sentences that express the associated relation and de-emphasize noisy sentences. The representation of a sentence x_i is obtained by concatenating the word and position embeddings [25].

Suppose a set S contains n sentences for an entity pair (e_1, e_2). Then, the set vector s is computed as the weighted sum of the sentence vectors x_i:

$$s = \sum_i \alpha_i x_i$$

$$\alpha_i = \frac{exp(x_i A r)}{\sum_k exp(x_k A r)}$$

where α_i is the weight of the sentence vector x_i, A is a weighted diagonal matrix and r is the query vector associated with the relation.

■ **Classification Layer:** The final classification layer employs the softmax loss function defined by Lin et al. [13]. The conditional probability $p(r|S, \theta)$ (θ denotes the model parameters) is given by:

$$p(r|S, \theta) = \frac{exp(o_r)}{\sum_{k=1}^{n_r} exp(o_k)}$$

where n_r is the total number of relations and o is the final output of the neural network model, which is given by:

$$o = Ms + d$$

where $d \in R^{n_r}$ is a bias vector and M is the representation matrix of relations.

4.   Experiments and Results

This section describes the experiments conducted to demonstrate that the proposed neural network model can effectively extract relations in unstructured threat intelligence. The held-out evaluation was employed and the aggregate precision/recall graphs are provided. The results reveal that the proposed neural network model has the best performance.

4.1   Experiment Details

This section provides details about the experiments, including the data sources, experimental dataset and parameter settings:

■ **Data Sources:** A web crawler based on the Scrapy framework was developed to harvest unstructured threat intelligence information

Table 1. Unstructured threat intelligence sources.

Source	Content	Format
FireEye	Security blog, security report	HTML, PDF
Symantec	Security blog, security report	HTML, PDF
Kaspersky	Security blog, security report	HTML, PDF
Cisco	Security blog	HTML
McAfee	Security blog	HTML
UNIT 42	Security blog	HTML
Twitter	Security tweet	HTML

from public security blogs and security reports released by prominent network security companies. In addition, the Twitter API was employed to harvest unstructured threat intelligence information from security practitioner tweets. Table 1 lists the threat intelligence sources. A total of 2,153 threat intelligence documents were collected.

Table 2. Top five relations in the dataset.

Relation	Instances
`/hackgroup/tool/use_tool`	1,160
`/hackgroup/organization/target_org`	1,140
`/hackgroup/location/target_loc`	1,094
`/hackgroup/method/texhnique/attack_method`	715
`/organization/hackgroup/investigate_hackgroup`	687
Total	4,796

- **Experimental Dataset:** Table 2 lists the top five relations extracted from the dataset. For each relation, 500 instances were randomly generated to produce the experimental dataset that contained 2,500 instances.

  Data associated with each relation was randomly partitioned into training (80%) and testing (20%) datasets. The training dataset contained 2,000 instances whereas the testing dataset contained 500 instances. The entire dataset is available at `github.com/luoluoluoyl/relation_extract_dataset.git`.

- **Parameter Settings:** Cross-validation of the training dataset was used to tune the neural network models. Table 3 shows all the

Table 3.   Parameter settings.

Parameter	Setting
Word embedding dimensions (d_w)	50
Position embedding dimensions (d_p)	5
Window size (l)	3
Batch size (b)	100
Maximum training iterations (max_{TI})	60
Learning rate (λ)	0.1
Dropout (p)	0.5

parameter settings. In particular, the number of word embedding dimensions d_w was set to 50, position embedding dimensions d_p to 5 and window size l to 3. The batch size B was set to 100. The maximum number of iterations for training max_{TI} was set to 60. The learning rate λ was set to 0.1 and dropout rate p to 0.5.

4.2   Comparison with Baseline Models

The neural network model developed in this research was compared with several state-of-the-art neural network models:

- **Bi-LSTM+ATT+NOPOS:** Zhou et al. [26] proposed the ATT-Bi-LSTM neural network model. Experimental results obtained for the SemEval-2010 relation classification task demonstrated the effectiveness of the ATT-Bi-LSTM model. Since the baseline ATT-Bi-LSTM model uses word embedding but not position embedding (NOPOS), it is named Bi-LSTM+ATT+NOPOS for comparison purposes.

- **Bi-LSTM+ONE:** Zeng et al. [25] proposed a neural network model based on the at-least-one assumption (ONE). The model incorporates a piecewise convolutional neural network with multi-instance learning for distant supervised relation extraction. In the experiments, the ONE assumption was applied in a Bi-LSTM model to create the Bi-LSTM+ONE baseline neural network model.

- **Bi-LSTM+CROSS MAX:** Jiang et al. [6] proposed a multi-instance multi-label neural network model for distant supervised relation extraction. It relaxes the at-least-once assumption (ONE) and uses cross-sentence max-pooling (CROSS MAX) to enable information sharing across different sentences. Overlapping relations are handled using multi-label learning with a neural network clas-

Table 4. Baseline neural network model results.

Neural Network Model	F1-Score	AUC	Accuracy
Word Embedding			
LSTM+ATT+NOPOS	0.5719	0.5991	0.7491
Bi-LSTM+ATT+NOPOS	0.6406	0.6695	0.8083
Word + Position Embedding			
LSTM+CROSS MAX	0.6641	0.7253	0.8750
LSTM+ONE	0.7047	0.7362	0.8674
LSTM+ATT	0.7312	0.7995	0.9253
Bi-LSTM+CROSS MAX	0.7069	0.7643	0.8977
Bi-LSTM+ONE	0.7730	0.8253	0.9300
Bi-LSTM+ATT	0.8207	0.9004	0.9784

sifier. In the experiments, CROSS MAX was combined with a Bi-LSTM model to create the Bi-LSTM+CROSS MAX baseline neural network model.

Cross-validation was performed on the three baseline neural network models. In order to compare the LSTM and Bi-LSTM models, four additional neural network models were specified as baselines: LSTM+ATT+NOPOS, LSTM+ATT, LSTM+CROSS MAX and LSTM+ONE.

Table 4 shows the F1-score, AUC (area under curve) and accuracy values for the baseline neural network models that only use word embedding and the baseline neural network models that use word and position embeddings. Figure 6 shows the precision/recall graphs obtained for all the baseline neural network models.

The experimental results motivate the following observations:

- The Bi-LSTM+ATT neural network model developed in this research significantly outperforms all the baseline neural network models with regard to relation extraction.

- As shown in Figures 6(a) and 6(b), the LSTM and Bi-LSTM neural network models that incorporate the ATT method have better performance than the LSTM and Bi-LSTM neural network models that incorporate the ONE and CROSS MAX methods.

- As shown in Figures 6(c), 6(d) and 6(e), a Bi-LSTM neural network model outperforms the baseline LSTM neural network models for all three methods (ATT, ONE and CROSS MAX).

- Table 4 shows that the LSTM+ATT+NOPOS and Bi-LSTM+ATT +NOPOS baseline neural network models that only use word em-

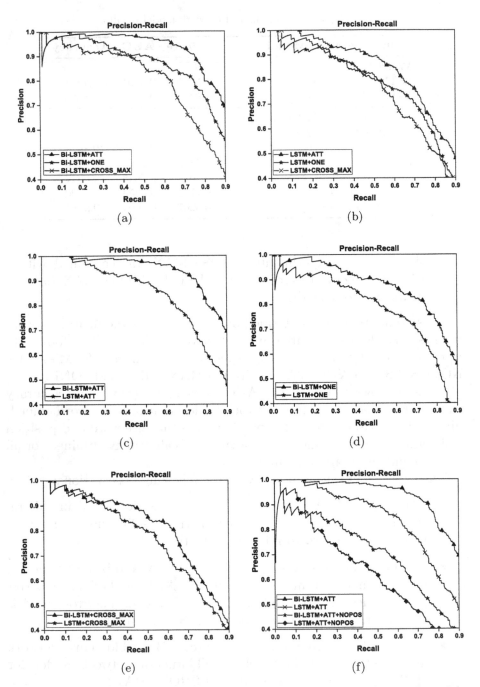

Figure 6.   Aggregate precision/recall graphs for the baseline neural network models.

Table 5. Analysis of the extraction results of four hacker groups.

Group	Tool Used	Target Org.	Target Location	Attack Method	Investig. Org.
Lazarus	KillDisk, PapaAlfa, Rising Sun, AIX	Sony Pictures Entertainment, South Korean Aerospace Companies	Africa, Europe, South Korea	Watering Hole, Phishing Email, Spear Phishing	McAfee. Kaspersky, Symantec, Novetta
Lucky Mouse	hTan, HyberPro	Data Center, Financial Services Company	Central Asia, Southeast Asia, America	Watering Hole, Phishing Email	Kaspersky, Palo Alto
APT10	Poison Ivy, PlugX, Quasar	Laoying Baichen Instruments Equipment Company, MSPs	Southeast Asia, United States, France, Germany	DLL Injection, Previous Credentials	FireEye, PwC UK, Recorded Future, BAE Systems
Nitro	Poison Ivy, Legitimate Compromised Websites	Chemical Company, Defense Company	South Korea	Spear Phishing, Remote Access	Symantec, Cyber Squared

bedding yield the worst results. This indicates that the position embedding proposed in this research is necessary and beneficial. This is also confirmed by the graphs in Figure 6(f).

4.3 Extraction Results

Five relations associated with hacker groups were extracted: Tool Used, Target Organization, Target Location, Attack Method and Investigating Organization. The extracted relations can be used to conduct a behavioral analysis of hacker groups. In addition to discovering the behavioral characteristics of hacker groups, it is possible to obtain correlations between the hacker groups.

Table 5 shows an analysis of four hacker groups. Using the extracted (head entity, tail entity, relation) tuples, it was possible to identify the tools used by the hacker groups, their targets and target locations, their attack methods and the organizations that investigated their attacks. For example, the Lazarus hacker group used a variety of attack methods (watering holes, phishing email and spear phishing) and, in addition to targets in South Korea, attacked targets in several countries in Africa and Europe.

The extracted information also enables commonalities between hacker groups to be discerned. For example, the Lazarus and Lucky Mouse hacker groups used watering holes and phishing email in their attacks, and the APT10 and Nitro hacker groups used the Poison Ivy tool to launch their attacks.

Such extraction results could be useful in an incident investigation. After an attack by an unknown hacker group is discovered, the tools and methods used in the attack would be determined during the investigation. Information about the tools and methods could then be compared with the extracted data. Hacking groups that match the tools and methods could be identified as suspects. In short, automatically extracting relationships from open-source intelligence helps build a knowledge base, which reduces the manual workload in investigations.

5. Conclusions

The automated extraction of relations from open-source threat intelligence can reduce the workload involved in security analyses and incident investigations. The framework described in this chapter employs distant supervision for data annotation and a Bi-LSTM neural network model to automatically extract threat intelligence relations. It effectively and efficiently alleviates the challenges involved in manual data annotation to create a high-quality labeled dataset for training neural network models. The Bi-LSTM neural network model used by the framework provides significant improvements in relation extraction performance over state-of-the-art neural network models.

Future research will create an open-source threat intelligence relation extraction benchmark dataset. It will also define more sophisticated relations and expand the annotation dataset. Efforts will also be made to further alleviate the incorrect data labeling problem and reduce the manual label verification workload.

Acknowledgement

This research was supported by the National Key Research and Development Program of China under Grant nos. 2016YFB0801004 and 2018YFC0824801.

References

[1] Amazon, Amazon Mechanical Turk, Seattle, Washington (`www.mturk.com`), 2021.

[2] S. Bird, NLTK: The Natural Language Toolkit, *Proceedings of the Twenty-First International Conference on Computational Linguistics and Forty-Fourth Annual Meeting of the Association for Computational Linguistics: Interactive Presentation Sessions*, pp. 69–72, 2006.

[3] H. Gascon, B. Grobauer, T. Schreck, L. Rist, D. Arp and K. Rieck, Mining attributed graphs for threat intelligence, *Proceedings of the Seventh ACM Conference on Data and Application Security and Privacy*, pp. 15–22, 2017.

[4] A. Graves and J. Schmidhuber, Framewise phoneme classification with bidirectional LSTM and other neural network architectures, *Neural Networks*, vol. 18(5-6), pp. 602–610, 2005.

[5] S. Hochreiter and J. Schmidhuber, Long short-term memory, *Neural Computation*, vol. 9(8), pp. 1735–1780, 1997.

[6] X. Jiang, Q. Wang, P. Li and B. Wang, Relation extraction with multi-instance multi-label convolutional neural networks, *Proceedings of the Twenty-Sixth International Conference on Computational Linguistics: Technical Papers*, pp. 1471–1480, 2016.

[7] C. Jones, R. Bridges, K. Huffer and J. Goodall, Towards a relation extraction framework for cyber security concepts, *Proceedings of the Tenth Annual Cyber and Information Security Research Conference*, article no. 11, 2015.

[8] B. Jordan and J. Wunder (Eds.), STIX 2.0 Specification, Core Concepts, Version 2.0 Draft 1, OASIS Cyber Threat Intelligence Technical Committee (`www.oasis-open.org/committees/download.php/58538/STIX2.0-Draft1-Core.pdf`), 2017.

[9] A. Joshi, R. Lal, T. Finin and A. Joshi, Extracting cybersecurity-related linked data from text, *Proceedings of the Seventh IEEE International Conference on Semantic Computing*, pp. 252–259, 2013.

[10] R. Lai, Information Extraction of Security-Related Terms and Concepts from Unstructured Text, M.S. Thesis, Department of Computer Science and Electrical Engineering, University of Maryland Baltimore County, Baltimore, Maryland, 2013.

[11] K. Lee, C. Hsieh, L. Wei, C. Mao, J. Dai and Y. Kuang, Sec-Buzzer: Cyber security emerging topic mining with open threat intelligence retrieval and timeline event annotation, *Soft Computing*, vol. 21(11), pp. 2883–2896, 2017.

[12] X. Liao, K. Yuan, X. Wang, Z. Li, L. Xing and R. Beyah, Acing the IOC game: Toward automatic discovery and analysis of open-source cyber threat intelligence, *Proceedings of the ACM SIGSAC Conference on Computer and Communications Security*, pp. 755–766, 2016.

[13] Y. Lin, S. Shen, Z. Liu, H. Luan and M. Sun, Neural relation extraction with selective attention over instances, *Proceedings of the Fifty-Fourth Annual Meeting of the Association for Computational Linguistics*, pp. 2124–2133, 2016.

[14] C. Manning, M. Surdeanu, J. Bauer, J. Finkel, S. Bethard and D. McClosky, The Stanford CoreNLP Natural Language Processing Toolkit, *Proceedings of the Fifty-Second Annual Meeting of the Association for Computational Linguistics: System Demonstrations*, pp. 55–60, 2014.

[15] R. McMillan, Definition: Threat Intelligence, Gartner, Stamford, Connecticut, 2013.

[16] N. McNeil, R. Bridges, M. Iannacone, B. Czejdo, N. Perez and J. Goodall, PACE: Pattern accurate computationally efficient bootstrapping for timely discovery of cyber security concepts, *Proceedings of the Twelfth International Conference on Machine Learning and Applications*, pp. 60–65, 2013.

[17] T. Mikolov, K. Chen, G. Corrado and J. Dean, Efficient estimation of word representations in vector space, presented at the *First International Conference on Learning Representations*, 2013.

[18] M. Mintz, S. Bills, R. Snow and D. Jurafsky, Distant supervision for relation extraction without labeled data, *Proceedings of the Joint Conference of the Forty-Seventh Annual Meeting of the Association for Computational Linguistics and Fourth International Joint Conference on Natural Language Processing of the Asian Federation of National Language Processing*, pp. 1003–1011, 2009.

[19] S. Mittal, P. Das, V. Mulwad, A. Joshi and T. Finin, CyberTwitter: Using Twitter to generate alerts for cyber security threats and vulnerabilities, *Proceedings of the IEEE/ACM International Conference on Advances in Social Networks Analysis and Mining*, pp. 860–867, 2016.

[20] V. Mulwad, W. Li, A. Joshi, T. Finin and K. Viswanathan, Extracting information about security vulnerabilities from web text, *Proceedings of the IEEE/WIC/ACM International Conferences on Web Intelligence and Intelligent Agent Technology*, pp. 257–260, 2011.

[21] J. Smith, Cyber threat intelligence sharing – Ascending the pyramid of pain, *APNIC Blog*, June 23, 2016.

[22] spaCy, spaCy – Industrial-Strength Natural Language Processing in Python (`spacy.io`), 2021.

[23] R. Steele, Open-source intelligence, in *Handbook of Intelligence Studies*, L. Johnson (Ed.), Routledge, Abingdon, United Kingdom, pp. 129–147, 2007.

[24] Y. Tao, Y. Zhang, S. Ma, K. Fan, M. Li, F. Guo and Z. Xu, Combining big data analysis and threat intelligence technologies for the classified protection model, *Cluster Computing*, vol. 20(2), pp. 1035–1046, 2017.

[25] D. Zeng, K. Liu, S. Lai, G. Zhou and J. Zhao, Relation classification via convolutional deep neural networks, *Proceedings of the Twenty-Fifth International Conference on Computational Linguistics: Technical Papers*, pp. 2335–2344, 2014.

[26] P. Zhou, W. Shi, J. Tian, Z. Qi, B. Li, H. Hao and B. Xu, Attention-based bidirectional long short-term memory networks for relation classification, *Proceedings of the Fifty-Fourth Annual Meeting of the Association for Computational Linguistics: Volume 2 Short Papers*, pp. 207–212, 2016.

Chapter 11

SECURITY AUDITING OF INTERNET OF THINGS DEVICES IN A SMART HOME

Suryadipta Majumdar, Daniel Bastos and Anoop Singhal

Abstract   Attacks on the Internet of Things are increasing. Unfortunately, transparency and accountability that are paramount to securing Internet of Things devices are either missing or implemented in a questionable manner. Security auditing is a promising solution that has been applied with success in other domains. However, security auditing of Internet of Things devices is challenging because the high-level security recommendations provided by standards and best practices are not readily applicable to auditing low-level device data such as sensor readings, logs and configurations. Additionally, the heterogeneous nature of Internet of Things devices and their resource constraints increase the complexity of the auditing process. Therefore, enabling the security auditing of Internet of Things devices requires the definition of actionable security policies, collection and processing of audit data, and specification of appropriate audit procedures.

This chapter focuses on the security auditing of Internet of Things devices. It presents a methodology for extracting actionable security rules from existing security standards and best practices and conducting security audits of Internet of Things devices. The methodology is applied to devices in a smart home environment, and its efficiency and scalability are evaluated.

Keywords: Internet of Things, security auditing, formal verification

1. Introduction

The popularity of Internet of Things (IoT) devices is growing rapidly. In fact, Statista [38] projects that more than 75.44 billion devices will be operational by 2025. However, current Internet of Things devices are easy targets of compromise due to implementation flaws and misconfigurations [1, 31, 43]. Verifying the flaws and misconfigurations, and

© IFIP International Federation for Information Processing 2021
Published by Springer Nature Switzerland AG 2021
G. Peterson and S. Shenoi (Eds.): Advances in Digital Forensics XVII, IFIP AICT 612, pp. 213–234, 2021.
https://doi.org/10.1007/978-3-030-88381-2_11

ensuring the accountability and transparency of the devices [1, 11] are essential for consumers and vendors.

Security auditing – verifying system states against a set of security rules – has become standard practice in enterprise security environments (see, e.g., Delloite [13], KPMG [23] and IBM [21]). Its advantages include supporting a range of security rules that cover system and network configurations, enabling examinations of the effects of events on system states, and delivering rigorous results via formal verification methods [25] as opposed to using other security solutions such as intrusion detection.

Security auditing has the potential to become a leading security measure against emerging threats to Internet of Things devices. However, security audits of Internet of Things devices are hindered by three principal challenges. First, existing security standards and best practices (e.g., [11, 14, 16, 17, 34]) provide high-level recommendations instead of guidance for conducting security audits of low-level system data in Internet of Things devices. Second, most Internet of Things devices cannot conduct the auditing process autonomously because they have limited resources to log audit data [43] and execute the formal verification tools used for auditing. Third, identifying the essential audit data to be collected for each security rule can be tedious.

Considerable research has focused on Internet of Things device security, including application monitoring, intrusion detection, device fingerprinting and access control. A few solutions provide *ad hoc* lists of rules for security tasks such as mobile application verification, network traffic inspection and access control [5, 6, 9, 43]. However, what is missing is a generic approach for automatically defining actionable rules that can be used to verify Internet of Things device security. Also missing is an auditing methodology designed for Internet of Things devices that can verify a range of security rules and present detailed audit reports with evidence of breaches. Unfortunately, existing auditing solutions for other environments such as the cloud (e.g., [10, 25, 26, 29]) are not applicable to Internet of Things devices because of their heterogeneous audit data sources, resource constraints and limited logging functionality.

This chapter proposes a security auditing methodology for Internet of Things devices. The methodology supports the extraction of actionable security rules from existing security standards and best practices, and enables security audits of Internet of Things devices. As a proof of concept, the security auditing methodology is applied to devices in a smart home environment, and its efficiency and scalability are evaluated (e.g., auditing 1,000 smart home networks within ten minutes).

2. Preliminaries

This section reviews key Internet of Things security standards and best practices, highlights the challenges to security auditing of Internet of Things devices, and defines the threat model.

2.1 Security Standards and Best Practices

This section reviews several security standards and best practices, namely NISTIR 8228 [11], NISTIR 8259 [17], OWASP IoT Security Guidance [34], ENISA Good Practices for Security of IoT [16] and the U.K. Government's Code of Practice for Consumer IoT Security [14]:

- **NISTIR 8228 [11]:** NIST's internal report NISTIR 8228 provides security and privacy recommendations for Internet of Things environments. It identifies four capabilities of Internet of Things devices: (i) transducer, (ii) data, (iii) interface and (iv) support. Additionally, it discusses a number of generic security recommendations for Internet of Things devices. In Section 4 below, security rules are specified based on the four device capabilities identified by NISTIR 8228.

- **NISTIR 8259 [17]:** Unlike NISTIR 8228, NIST's internal report NISTIR 8259 provides specific security recommendations for Internet of Things device vendors. It identifies six activities that vendors should consider during the pre-market and post-market phases. Additionally, it identifies five risk mitigation goals for consumers: (i) asset management, (ii) vulnerability management, (iii) access management, (iv) data protection and (v) incident detection.

- **OWASP IoT Security Guidance [34]:** OWASP's IoT Security Guidance includes recommendations for device vendors, application developers and consumers. The recommendations, which are divided into ten categories, are used in Section 4 to identify actionable security rules for Internet of Things devices.

- **ENISA Good Practices for Security of IoT [16]:** ENISA's Good Practices for Security of IoT focuses on the software development lifecycle. Its principal audiences are Internet of Things software developers, integrators, and platform and system engineers. It provides recommendations for preventing the introduction of vulnerabilities via the insecure implementation of software development lifecycle processes. The main contributions are an analysis of security concerns in all phases of the software development life-cycle, detailed asset and threat taxonomies, good practices that enhance security and their mappings to related standards, guidelines and schemes.

■ **U.K. Government Code of Practice [14]:** The code of practice published by the Department for Digital, Culture, Media and Sport of the U.K. Government focuses on Internet of Things device security. The code of practice includes recommendations for device vendors, application developers and service providers. The recommendations, which are divided into 13 categories, are used in Section 4 to identify actionable security rules for Internet of Things devices.

2.2   Security Auditing Challenges

Security auditing of Internet of Things devices faces four principal challenges:

■ Security standards and best practices such as NIST 8228 [11] and OWASP IoT Security Guidance [34] provide high-level recommendations to programmers and practitioners instead of guidance for conducting automated security operations such as monitoring and auditing Internet of Things devices. As a result, the recommendations cannot be used directly to create actionable rules for verifying Internet of Things device security. For instance, the OWASP high-level recommendation "ensure proper authentication/authorization mechanisms" must be instantiated to an actionable rule such as "require a unique username and complex password of more than eight characters to access a smart door" in order to enable security verification.

■ The recommendations in NISTIR 8228 [11], NISTIR 8259 [17], OWASP IoT Security Guidance [34], ENISA Good Practices for Security of IoT [16] and U.K. Government Code of Practice [14] differ significantly in their scopes, objectives and descriptions. Furthermore, some recommendations conflict with each other. A single comprehensive source for actionable rules does not exist. As a result, it is necessary to systematically analyze all the high-level recommendations, interpret them, resolve conflicts and then derive actionable security rules.

■ System information such as the hardware specifications and software APIs of Internet of Things devices from different vendors are published in different formats and use vendor-specific jargon [15]. Therefore, it is essential to first normalize the vendor-specific materials and interpret the high-level recommendations specified in the standards and best practices in the context of the various implementations.

- Even after actionable rules are specified, Internet of Things devices are unable to conduct auditing processes autonomously because they have limited storage for logging audit data and computational power for executing formal verification tools. For example, a smart door would not be able to execute Sugar [40], a Boolean satisfiability solver.

This work attempts to overcome these challenges by deriving actionable rules for verifying Internet of Things device security and designing a security auditing methodology for Internet of Things devices.

2.3 Threat Model

The threat model assumes that Internet of Things devices have implementation flaws, misconfigurations and vulnerabilities that are exploited by malicious entities to violate security rules. A remote server or a local hub or gateway is required to perform security audits. The communications between Internet of Things devices and the verification server are secured by end-to-end encryption mechanisms. Privacy threats associated with data sharing by Internet of Things devices are beyond the scope of this research. However, they could be addressed in future work using a privacy-friendly auditing technique.

To keep the discussion concrete, the remainder of this work uses a smart home environment to elaborate the concepts underlying the proposed security auditing methodology.

3. Security Auditing Methodology

Figure 1 provides an overview of the security auditing methodology for Internet of Things devices. The methodology involves three steps:

- **Step 1:** Build a knowledge base from Internet of Things device security standards and best practices, and details of Internet of Things device designs and implementations.

- **Step 2:** Translate the system knowledge and keywords to security rules.

- **Step 3:** Audit Internet of Things devices using the security rules.

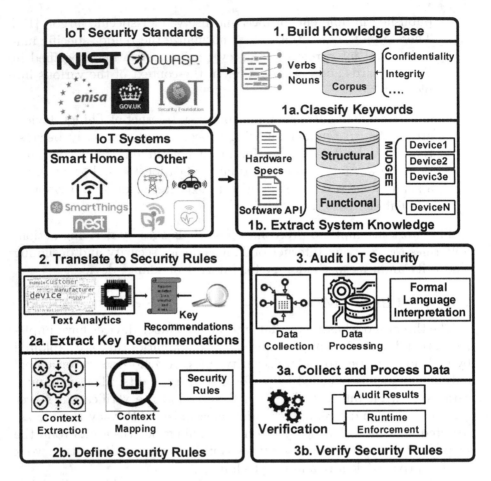

Figure 1.   Security auditing methodology.

3.1   Step 1: Build a Knowledge Base

In order to audit the security of Internet of Things devices in a smart home, it is important to understand existing security standards and best practices as well as details of Internet of Things device designs and implementations. A knowledge base is created to codify this knowledge. Creating the knowledge base involves two steps:

- *Extract Keywords from the Guidelines Listed in Security Standards and Best Practices:* Keywords are extracted by parsing the contents of relevant sections in security standards and best practices documents. A corpus is then created with the relevant terms, mainly nouns and verbs, the two parts of speech that convey the essence of the recommendations. Finally, the keywords in the corpus are classified based on standard security goals such as confidentiality, integrity and availability [3].

- *Collect System Knowledge Related to Internet of Things Device Designs and Implementations:* Structural knowledge about devices such as sensors and actuators is obtained from vendor-provided materials, including hardware specifications and software APIs as described in [15]. Functional knowledge such as network protocol usage is obtained by analyzing the network behavior of Internet of Things devices using MUDGEE [19].

3.2 Step 2: Translate to Security Rules

Having created the knowledge base, knowledge in the repository is translated to actionable security rules for Internet of Things devices. The translation process involves two steps:

- *Extract Recommendations from the Classified Keywords in the Security Standards and Best Practices:* The extraction of recommendations is automated using text analysis algorithms such as term frequency-inverse document frequency and sentiment analysis [39]. Next, the results are manual inspected to create a shortlist of the recommendations.

- *Define Actionable Security Rules by Instantiating the Recommendations with System Knowledge:* Actionable rules are defined by extracting the contexts of the recommendations using deep contextualized learning [35]. The context associated with each recommendation is mapped to related system knowledge and each recommendation is then mapped to a concrete security rule.

3.3 Step 3: Audit IoT Device Security

Having obtained a set of actionable security rules, the final objective is to conduct security audits of Internet of Things devices. Security auditing involves two steps:

- *Collect and Process Audit Data for the Security Rules Covering Internet of Things Devices:* The audit data sources corresponding to the security rules for each Internet of Things device are identified and the logged data is collected. The collected data is converted to a formal language format such as first-order logic.

- *Verify the Security Rules Using Formal Verification:* The first-order logic expressions are converted to the input format required by a formal verification tool such as Sugar [40], a Boolean satisfiability solver. The verification results are interpreted. Finally the auditing results are presented to support various capabilities such as evidence analysis and security decision enforcement (e.g., allow or deny at runtime).

4.   Auditing Smart Home Security

This section demonstrates the application of the proposed methodology in a use case involving the security auditing of Internet of Things devices in a smart home environment.

4.1   Security Rule Definition

This first step in the security auditing methodology is to define concrete security rules. To establish a bridge between the high-level guidelines in security standards and best practices and low-level system information pertaining to smart home devices, concrete security rules from the standards and best practices and the literature were specified to formulate the security auditing problem.

Table 1 shows sample security auditing rules identified from relevant standards and best practices, smart home literature and real-world smart home implementations (e.g., Google Nest). The specific standards and best practices used were NIST 8228 [11], OWASP IoT Security Guidance [34] and U.K. Government Code of Practice [14].

The running example in this chapter will consider the following versions of rules *R1* and *R6* in Table 1:

- R1: Smart lock must not be in the unlocked state when other devices are in the vacation mode.

- R6: Photo and video captures are not allowed in a bathroom.

4.2   Data Collection

The next step is to collect the audit data to verify the security rules. Figure 2 shows sample data collected about Google Nest products. To obtain the data to verify the cross-device rules, the fields noted in the blueprints must be extracted from the event logs corresponding to each rule. Note that the required data was already collected by the devices, so no changes to the devices were necessary. In order to obtain data for auditing the rules pertaining to device capabilities (rules *R7* and *R8*), the types of all the installed devices were identified by analyzing the network traffic using the IoT Inspector [20] and IoTSense [7] tools.

Table 1. Sample security rules for conducting security audits of devices in a smart home.

Security Rule	Description	Standards and Best Practices		
		NIST8228 [11]	OWASP [34]	UK Govt. [14]
R1: No unauthorized access 1	Smart lock must be in the locked state when the smart plug is in the vacation state [6, 43]	DE.AE-3, DE.CM-7	I6, I8, I9	6, 10
R2: No unauthorized access 2	Smart lock must be in the locked state when the smart light must be in the off state [6, 43]	DE.AE-3, DE.CM-7	I6, I8, I9	6, 10
R3: No lighting in an empty home	Smart light must be in the off state when the smart plug is in the vacation state [6, 36]	DE.AE-3	I6, I7, I9	6, 7, 10
R4: No sound in the sleep state	Speaker must be in the silent state when the clock is in the sleep state [6]	DE.AE-3	I6, I9	6, 7, 10
R5: Consistent climate control	Temperature cannot simultaneously correspond to cooling and heating [6]	DE.AE-3	I9	6, 7, 10
R6: No sensitive information leak	Device location must not be sensitive if the device has a camera [6]	DE.CM-7	I5	*6, 7, 10*
R7: Proper flood detection	Water flooding alarm is in the on state when the water level is higher than the threshold [6]	PR.AC-4	I6, I7, I9	7, 10
R8: Proper smoke detection	Smoke alarm state is in the on state when the smoke amount is higher than the threshold [42]	PR.AC-4	I6, I7, I9	7, 10

Table 2.   Sample data collected about Google Nest products.

Device	Collected Data
Smoke Detector	id:1, device_id:vgUlapP6, locale:en-US, software_version:4.0, last_connection: 2018-12-31T23:59:59.000Z, battery_health:replace, co_alarm_state:ok, smoke_alarm_state:ok
Camera	id:1, device_id:2saNS6kQ, software_version:3.9, name:Front Door, is_streaming:false, web_url:https://home.nest.com/cameras/2saNS6kQ, is_online:false
Thermostat	id:4, device_id:vgUlapP6, locale:en-US, software_version:3.6, last_connection: 2019-1-05T15:59:59.000Z, ambient_temperature_f:70, ambient_temperature_c:38, humidity:75

Example 1: In the case of rule *R1*, event logs of the smart lock, smart plug and thermostat were collected. The following data was collected: smart_lock1.lock_state:locked, smart_lock2.lock_state:unlocked, smart_plug1.vacation_state:on, thermostat1.vacation_state:on.

In the case of rule *R6*, the sensing capabilities of all the bathroom devices were collected: showerhead:{bluetooth, microphone}, smart mirror:{camera, ambient light sensor, motion sensor}, water pebble:{bluetooth}.

4.3   Formal Language Translation

The next step is to convert the audit data and security rules to formal language expressions for the verification step. To this end, the security rules were expressed in the input format of the formal verification tool, i.e., as a constraint satisfaction problem for Sugar [40]. Programs were developed to translate the collected audit data for input to the selected verification tool.

Example 2: Rule R1 is expressed as the predicate:

$$\forall \mathtt{l} \in \mathtt{Smartlock},\ \forall \mathtt{p} \in \mathtt{Smartplug},\ \forall \mathtt{h} \in \mathtt{Smarthome},\ \forall \mathtt{s} \in \mathtt{DeviceStatus}$$
$$(\mathtt{LockState(h,l,l.s)} \wedge \mathtt{VacationState(h,p,p.s)}) \wedge (\mathtt{IsOff(l.s)}) \longrightarrow$$
$$(\mathtt{IsOff(p.s)})$$

The corresponding constraint satisfaction problem (CSP) predicate is:

```
(and LockState(h,l,l.s) VacationState(h,p,p.s) (IsOff(l.s))
                            (not (IsOff(p.s)))
```

The `LockState(h,l,s)` relation indicates the lock status `s` of smart lock `l` in smart home `h`. The `VacationState(h,p,s)` relation indicates the vacation status `s` of smart plug `p` in smart home `h`. The `IsOff(l.s)` relation indicates that the status of device `d` is `OFF`. For example, the `lockState(h1,l1,s1)` relation is `TRUE` when there exists a smart home `h1` with smart lock `l1` in state `s1`; otherwise the relation is `FALSE`. Similarly, the other relations in the CSP predicate are evaluated. Note that the CSP predicate becomes `TRUE` when rule *R1* is breached. Next, the relations are instantiated for each tuple associated with the audit data as: `(LockState (supports (h1,l1,OFF) (h1,l2,ON) ...))` and `(VacationState (supports (h1,p1,ON) ...))`.

4.4   Verification

Verification leverages formal techniques such as Boolean satisfiability (SAT), declarative logic programming (Datalog) and satisfiability modulo theory (SMT), which have been used in several security applications [10, 24, 27, 28, 33]. These techniques are recognized for their expressivity of security rules, provable security and rigorous results. The verification tools are hosted at a server to overcome the resource constraints of Internet of Things devices. Note that no changes to the design or functionality of the verification tools are required.

Example 3: The CSP predicate of rule *R1* instantiated with the states of smart locks and smart plugs is executed in Sugar [40], a satisfiability problem solver.

4.5   Evidence Extraction

The final step is to interpret the outcome of the formal verification and prepare the audit reports. This effort is very specific to the verification tools that are used because most formal tools have their own output formats. However, in all cases, they provide rigorous results to identify evidence of any and all security rule violations.

Example 4: After verifying the CSP predicate of rule *R1* using the collected data, Sugar returns `SAT`, which indicates that the predicate is `TRUE` for the given data and that rule *R1* is violated. Furthermore, as evidence, Sugar identifies the tuple that caused the breach. Specifically, `LockState(h1,l1,OFF)` and `VacationState(h1,p1,ON)`, which mean

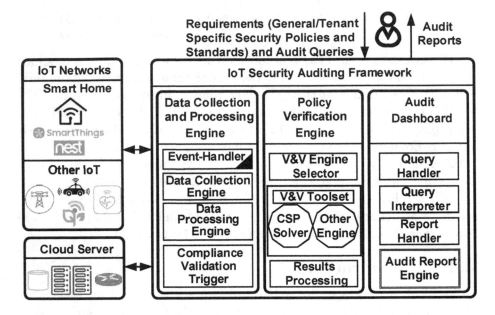

Figure 2.   Security auditing framework.

that smart lock **l1** is unlocked when smart plug **p1** is in the vacation state in the same smart home **h1**.

5.   Security Auditing Framework

Figure 2 shows a high-level representation of the security auditing framework for Internet of Things devices. The framework primarily interacts with Internet of Things devices to collect audit data and with a cloud server to store audit data and delegate auditing computations (i.e., verification). It also interacts with users such as security experts and smart home owners to obtain the auditing requirements (security policies) and provide them with audit results in the form of reports. The framework is designed to work with a variety of Internet of Things networks. However, this work is restricted to smart home environments.

The framework has three main components: (i) data collection and processing engine, (ii) policy verification engine and (iii) audit dashboard:

- ■ **Data Collection and Processing Engine:** The data collection and processing engine incorporates data collection and data processing (sub)engines. The data collection engine collects the required audit data in the batch mode using smart home platforms such as Google Nest. Required audit data may also be collected from an Internet of Things hub and/or Internet of Things cloud server depending on the specific smart home implementation.

The data processing engine filters the collected data to retain the data needed to verify the security rules. It converts the data to a uniform format and subsequently translates it to formal language expressions. The final processing steps generate the code for verification and store it in the audit repository database for use by the policy verification engine. The code that is generated depends on the verification engine that is employed.

- **Policy Verification Engine:** The policy verification engine verifies security policies and identifies security violations. Upon receiving an audit request and/or updated inputs, the engine invokes the back-end verification and validation (V&V) algorithms. Formal methods are employed to express system models and audit policies, facilitating automated reasoning, which is more practical and effective than manual analysis.

  When a security audit policy is breached, supporting evidence is obtained from the output of the verification back-end. After the compliance validation is completed, the audit results and evidence are stored in the audit repository database and made accessible to the audit report engine. Depending on the security policies being verified, multiple formal verification engines may be incorporated.

- **Auditing Dashboard:** The auditing dashboard enables users to select various standards and best practices as well as specific security policies drawn from the standards and best practices. After an auditing request is submitted and processed, summarized results are presented via the auditing dashboard. Details of the violations are provided along with their supporting evidence. Audit reports are archived for stipulated periods of time.

6.   Experiments and Results

This section presents the proof-of-concept experiments and their results.

6.1   Experimental Setup

The experiments employed physical and virtual testbeds. The physical testbed comprised 23 smart home products from several vendors,

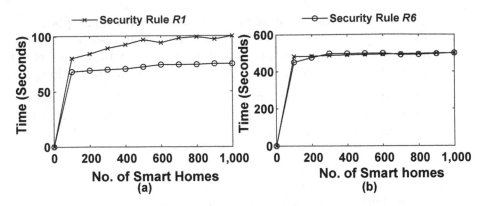

Figure 3. Time efficiency of the security auditing methodology.

15 Raspberry Pi single-board computers and 11 sensors for configuring the smart home devices. The virtual testbed simulated five smart home products: smart lock, smart plug, thermostat, camera and smoke detector.

The devices in the two testbeds produced outputs in standard formats (e.g., based on Google Nest API specifications [18]) and transmitted them to a cloud server for storage in a MySQL database. The two testbeds were connected to a virtual hub configured on a cloud server (Microsoft Azure IoT Hub [30]) or a physical hub. The hubs were connected to an auditing server with a 3.70 GHz Intel Core i7 Hexa core CPU and 32 GB memory.

The two testbeds were employed to generate datasets for the experiments. First, the physical testbed generated real data from the smart home products. The virtual testbed generated scaled-up data for up to 1,000 smart home networks based on the real data to help evaluate the scalability of the security auditing methodology. Each experiment was repeated 100 times and the average measurements were employed in the evaluations.

6.2   Experimental Results

The first set of experiments sought to measure the time efficiency of the security auditing methodology. Figure 3 shows the total times required to individually verify rule *R1* (which mandates that no unauthorized door be opened) and rule *R6* (which mandates that photo and video captures are not allowed in a bathroom) for up to 1,000 smart homes. Figure 3(a) shows the total verification times for five devices per smart home whereas Figure 3(b) shows the total verification times for 15 devices per smart home. The results reveal that the total times are

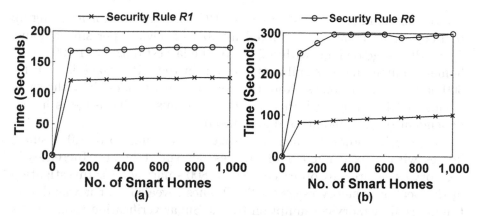

Figure 4. Data collection and processing time reqiurements.

not linear functions of the number of smart homes to be verified. Additional results (not discussed here due to space constraints) reveal that auditing additional security rules would not lead to significant increases in the total verification times.

Figure 4 shows the data collection and data processing times for rules *R1* and *R6* for up to 1,000 smart homes. Figure 4(a) shows the total times for five devices per smart home whereas Figure 4(b) shows the total times for 15 devices per smart home. Since the data collection and processing tasks are each performed only once for each audit request, the overheads are acceptable for auditing such large environments.

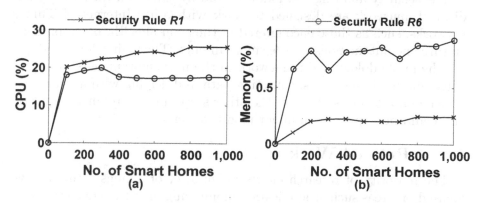

Figure 5. CPU and memory usage.

The second set of experiments sought to measure the CPU and memory usage of the security auditing methodology. Figure 5 shows the CPU and memory usage results for auditing rules *R1* and *R6* for up to 1,000 smart homes. Figure 5(a) shows the CPU usage results for auditing up to 1,000 smart homes with a fixed number of devices per smart home. The CPU usage remains within 26% for the largest dataset (1,000 smart homes). Furthermore, significant leveling in the CPU usage is seen for 300 or more smart homes. Note that other security rules show the same trends in CPU usage, which is expected because CPU usage is highly dependent on the amount of data collected.

Figure 5(b) shows the memory usage for auditing up to 1,000 smart homes with a fixed number of devices per smart home. Increases in memory usage are only observed beyond 800 smart homes. Investigation of the peak in memory usage for 200 homes revealed that it was due to the internal memory consumption by the Sugar verification tool. In any case, the highest memory usage over all the experiments is just 0.92%. Other security rules show the same trends in memory usage due to the high dependence of memory usage on the amount of data collected.

7.   Discussion

Due to the expressiveness of the underlying formal verification method (i.e., SAT solver), the security auditing methodology can support a wide-range of security rules. Specifically, any rule that can be expressed as a constraint satisfaction problem would be supported. The main overhead in adding new rules comes from identifying the data required for auditing and locating their sources. This support can be provided by integrating additional tools in the security auditing framework.

The security auditing methodology and by extension the security auditing framework are designed to work with other Internet of Things networks, such as those encountered in the smart health, precision agriculture and autonomous vehicle environments. The main effort in adapting the methodology and framework to the new environments would involve finding data sources, collecting data and dealing with application-specific data formats. Most of the other steps are environment-agnostic and could be applied with minor modifications.

8.   Related Work

The majority of research efforts in Internet of Things security have focused on areas such as application monitoring, intrusion detection and access control [2, 6, 8, 12, 32, 37, 43, 44, 46]. This section attempts to show that, while the research described in this chapter differs from other Internet of Things security approaches in terms of scope and objectives, they complement each other.

Application monitoring techniques execute the source code of Internet of Things device applications to analyze the applications. For example, ContextIoT [22] and SmartAuth [41] are permission-based systems for monitoring individual applications. ProvThings [43] creates provenance graphs using security-critical APIs to support Internet of Things device forensics. Soteria [4] and IoTGuard [6] verify security and safety policies by performing static and dynamic code analyses, respectively. Unlike these source code analysis tools, the research described in this chapter presents a concrete solution for conducting security audits of the logs and configurations of Internet of Things devices.

Several security solutions have been developed for smart homes. For example, Zhang et al. [46] monitor isolation-related properties of Internet of Things devices using a virtual channel. Yang et al. [44] protect Internet of Things devices by hiding them in onion gateways. HoMonit [45] proposes a smart home monitoring system that employs deterministic finite automaton models for Internet of Things devices. However, none of these methods offer a security auditing solution for Internet of Things devices.

Modern security auditing methodologies can be categorized as retroactive, incremental or proactive. A retroactive approach [25] audits a system after the fact, which means that it does not prevent irreversible damage such as denial of service or sensitive data loss. An incremental auditing approach [10, 28] audits the impacts of a change event; whereas such an approach overcomes the limitations of a retroactive approach, it causes significant delays in the response time. A proactive approach [10, 26, 29] computes a portion of the auditing effort in advance to keep runtime computations lightweight and, thus, provides practical response times, e.g., a few milliseconds to audit a mid-sized cloud [26]. However, unlike the security auditing methodology presented in this chapter, these auditing methods are not applicable to Internet of Things devices. This is primarily due to the computational and storage constraints of Internet of Things devices, their heterogeneity and limited logging functionality.

9. Conclusions

The proposed security auditing methodology for Internet of Things devices facilitates the extraction of actionable security rules from existing security standards and enables the automated auditing of the security rules using formal verification techniques and tools. Experiments conducted on physical and virtual testbeds with Internet of Things devices in smart home environments demonstrate the efficiency and scalability of the security auditing methodology, including the ability to complete security audits of 1,000 smart home networks within ten minutes.

Future research will focus on injecting privacy into the security auditing process, which is required because security policy verification currently uses a remote server. Research will also focus on automating the important task of converting raw information from security standards and best practices into actionable security rules, which is currently performed manually. Other research will consider simplifying the auditing workload using an incremental as opposed to a batch approach. Future research will also investigate applications of the security auditing methodology and by extension the security auditing framework to other Internet of Things networks, such as those encountered in smart health, precision agriculture and autonomous vehicle environments.

This chapter is not subject to copyright in the United States. Commercial products are identified in order to adequately specify certain procedures. In no case does such an identification imply a recommendation or endorsement by the National Institute of Standards and Technology, nor does it imply that the identified products are necessarily the best available for the purpose.

References

[1] O. Alrawi, C. Lever, M. Antonakakis and F. Monrose, SoK: Security evaluation of home-based IoT deployments, *Proceedings of the IEEE Symposium on Security and Privacy*, pp. 1362–1380, 2019.

[2] M. Balliu, M. Merro and M. Pasqua, Securing cross-app interactions in IoT platforms, *Proceedings of the Thirty-Second IEEE Computer Security Foundations Symposium*, pp. 319–334, 2019.

[3] C. Bellman and P. van Oorschot, Best practices for IoT security: What does that even mean? arXiv: 2004.12179 (`arxiv.org/abs/2004.12179`), 2020.

[4] Z. Berkay Celik, P. McDaniel and G. Tan, Soteria: Automated IoT safety and security analysis, *Proceedings of the USENIX Annual Technical Conference*, pp. 147–158, 2018.

[5] Z. Berkay Celik, P. McDaniel, G. Tan, L. Babun and A. Selcuk Uluagac, Verifying Internet of Things safety and security in physical spaces, *IEEE Security and Privacy*, vol. 17(5), pp. 30–37, 2019.

[6] Z. Berkay Celik, G. Tan and P. McDaniel, IoTGuard: Dynamic enforcement of security and safety policy in commodity IoT, *Proceedings of the Network and Distributed Systems Security Symposium*, 2019.

[7] B. Bezawada, M. Bachani, J. Peterson, H. Shirazi, I. Ray and I. Ray, Behavioral fingerprinting of IoT devices, *Proceedings of the Workshop on Attacks and Solutions in Hardware Security*, pp. 41–50, 2018.

[8] S. Bhatt, F. Patwa and R. Sandhu, An access control framework for cloud-enabled wearable Internet of Things, *Proceedings of the Third IEEE International Conference on Collaboration and Internet Computing*, pp. 328–338, 2017.

[9] S. Birnbach, S. Eberz and I. Martinovic, Peeves: Physical event verification in smart homes, *Proceedings of the Twenty-Sixth ACM SIGSAC Conference on Computer and Communications Security*, pp. 1455–1467, 2019.

[10] S. Bleikertz, C. Vogel, T. Gross and S. Modersheim, Proactive security analysis of changes in virtualized infrastructure, *Proceedings of the Thirty-First Annual Computer Security Applications Conference*, pp. 51–60, 2015.

[11] K. Boeckl, M. Fagan, W. Fisher, N. Lefkovitz, K. Megas, E. Nadeau, D. Gabel O'Rourke, B. Piccarreta and K. Scarfone, Considerations for Managing Internet of Things (IoT) Cybersecurity and Privacy Risks, NISTIR 8228, National Institute of Standards and Technology, Gaithersburg, Maryland, 2019.

[12] J. Choi, H. Jeoung, J. Kim, Y. Ko, W. Jung, H. Kim and J. Kim, Detecting and identifying faulty IoT devices in smart homes with context extraction, *Proceedings of the Forty-Eighth Annual IEEE/IFIP International Conference on Dependable Systems and Networks*, pp. 610–621, 2018.

[13] Deloitte, Cybersecurity and the Role of Internal Audit, New York (www2.deloitte.com/us/en/pages/risk/articles/cyber security-internal-audit-role.html), 2019.

[14] Department for Digital, Culture, Media and Sport, Code of Practice for Consumer IoT Security, Government of the United Kingdom, London, United Kingdom, 2018.

[15] A. Dolan, I. Ray and S. Majumdar, Proactively extracting IoT device capabilities: An application to smart homes, *Proceedings of the Thirty-Fourth Annual IFIP WG 11.3 Conference on Data and Applications Security and Privacy*, pp. 42–63, 2020.

[16] European Union Agency for Cybersecurity, Good Practices for Security of IoT: Secure Software Development Lifecycle, Athens, Greece (www.enisa.europa.eu/publications/good-practices-for-security-of-iot-1), 2019.

[17] M. Fagan, K. Megas, K. Scarfone and M. Smith, Foundational Cybersecurity Activities for IoT Device Manufacturers, NISTIR 8259, National Institute of Standards and Technology, Gaithersburg, Maryland, 2020.

[18] Google, Nest API Reference, Mountain View, California (`deve lopers.nest.com/reference/api-overview`), 2019.

[19] A. Hamza, D. Ranathunga, H. Gharakheili, M. Roughan and V. Sivaraman, Clear as MUD: Generating, validating and applying IoT behavioral profiles, *Proceedings of the Workshop on IoT Security and Privacy*, pp. 8–14, 2018.

[20] D. Huang, N. Apthorpe, F. Li, G. Acar and N. Feamster, IoT Inspector: Crowdsourcing labeled network traffic from smart home devices at scale, *Proceedings of the ACM on Interactive, Mobile, Wearable and Ubiquitous Technologies*, vol. 4(2), article no. 46, 2020.

[21] IBM, IBM Cloud Compliance Program, Armonk, New York (`www.ibm.com/cloud/compliance`), 2021.

[22] Y. Jia, Q. Chen, S. Wang, A. Rahmati, E. Fernandes, Z. Mao and A. Prakash, ContexIoT: Towards providing contextual integrity to appified IoT platforms, *Proceedings of the Network and Distributed Systems Security Symposium*, 2017.

[23] KPMG, Governance, Risk and Compliance Services, New York (`home.kpmg/xx/en/home/services/advisory/risk-consulting /internal-audit-risk.html`), 2021.

[24] T. Madi, Y. Jarraya, A. Alimohammadifar, S. Majumdar, Y. Wang, M. Pourzandi, L. Wang and M. Debbabi, ISOTOP: Auditing virtual network isolation across cloud layers in OpenStack, *ACM Transactions on Privacy and Security*, vol. 22(1), article no. 1, 2018.

[25] T. Madi, S. Majumdar, Y. Wang, Y. Jarraya, M. Pourzandi and L. Wang, Auditing security compliance of the virtualized infrastructure in the cloud: Application to OpenStack, *Proceedings of the Sixth ACM Conference on Data and Application Security and Privacy*, pp. 195–206, 2016.

[26] S. Majumdar, Y. Jarraya, M. Oqaily, A. Alimohammadifar, M. Pourzandi, L. Wang and M. Debbabi, LeaPS: Learning-based proactive security auditing for clouds, *Proceedings of the Twenty-Second European Symposium on Research in Computer Security, Part II*, pp. 265–285, 2017.

[27] S. Majumdar, T. Madi, Y. Wang, Y. Jarraya, M. Pourzandi, L. Wang and M. Debbabi, Security compliance auditing of identity and access management in the cloud: Application to OpenStack, *Proceedings of the Seventh IEEE International Conference on Cloud Computing Technology and Science*, pp. 58–65, 2015.

[28] S. Majumdar, T. Madi, Y. Wang, Y. Jarraya, M. Pourzandi, L. Wang and M. Debbabi, User-level runtime security auditing for the cloud, *IEEE Transactions on Information Forensics and Security*, vol. 13(5), pp. 1185–1199, 2018.

[29] S. Majumdar, A. Tabiban, M. Mohammady, A. Oqaily, Y. Jarraya, M. Pourzandi, L. Wang and M. Debbabi, Proactivizer: Transforming existing verification tools into efficient solutions for runtime security enforcement, *Proceedings of the Twenty-Fourth European Symposium on Research in Computer Security, Part II*, pp. 239–262, 2019.

[30] Microsoft, Azure IoT Hub, Redmond, Washington (`azure.micro soft.com/en-ca/services/iot-hub`), 2019.

[31] S. Notra, M. Siddiqi, H. Gharakheili, V. Sivaraman and R. Boreli, An experimental study of security and privacy risks with emerging household appliances, *Proceedings of the IEEE Conference on Communications and Network Security*, pp. 79–84, 2014.

[32] T. O'Connor, R. Mohamed, M. Miettinen, W. Enck, B. Reaves and A. Sadeghi, Homesnitch: Behavior transparency and control for smart home IoT devices, *Proceedings of the Twelfth Conference on Security and Privacy in Wireless and Mobile Networks*, pp. 128–138, 2019.

[33] OpenStack Project, OpenStack Congress, Austin, Texas (`wiki.openstack.org/wiki/Congress`), 2015.

[34] OWASP Foundation, IoT Security Guidance, Bel Air, Maryland (`www.owasp.org/index.php/IoT_Security_Guidance`), 2019.

[35] M. Peters, M. Neumann, M. Iyyer, M. Gardner, C. Clark, K. Lee and L. Zettlemoyer, Deep contextualized word representations, *Proceedings of the Annual Conference of the North American Chapter of the Association for Computational Linguistics, Volume 1 (Long Papers)*, pp. 2227–2237, 2018.

[36] E. Ronen and A. Shamir, Extended functionality attacks on IoT devices: The case of smart lights, *Proceedings of the IEEE European Symposium on Security and Privacy*, pp. 3–12, 2016.

[37] M. Serror, M. Henze, S. Hack, M. Schuba and K. Wehrle, Towards in-network security for smart homes, *Proceedings of the Thirteenth International Conference on Availability, Reliability and Security,* article no. 18, 2018.

[38] Statista, Internet of Things (IoT) connected devices installed base worldwide from 2015 to 2025, New York (www.statista.com/statistics/471264/iot-number-of-connected-devices-worldwide), November 27, 2016.

[39] M. Taboada, J. Brooke, M. Tofiloski, K. Voll and M. Stede, Lexicon-based methods for sentiment analysis, *Computational Linguistics,* vol. 37(2), pp. 267–307, 2011.

[40] N. Tamura and M. Banbara, Sugar: A CSP to SAT translator based on order encoding, *Proceedings of the Second International CSP Solver Competition,* pp. 65–69, 2008.

[41] Y. Tian, N. Zhang, Y. Lin, X. Wang, B. Ur, X. Guo and P. Tague, SmartAuth: User-centered authorization for the Internet of Things, *Proceedings of the Twenty-Sixth USENIX Security Symposium,* pp. 361–378, 2017.

[42] P. Vervier and Y. Shen, Before toasters rise up: A view into the emerging IoT threat landscape, *Proceedings of the Twenty-First International Symposium on Research in Attacks, Intrusions and Defenses,* pp. 556–576, 2018.

[43] Q. Wang, W. Ul Hassan, A. Bates and C. Gunter, Fear and logging in the Internet of Things, *Proceedings of the Network and Distributed Systems Security Symposium,* 2018.

[44] L. Yang, C. Seasholtz, B. Luo and F. Li, Hide your hackable smart home from remote attacks: The multipath onion IoT gateways, *Proceedings of the Twenty-Third European Symposium on Research in Computer Security, Part I,* pp. 575–594, 2018.

[45] W. Zhang, Y. Meng, Y. Liu, X. Zhang, Y. Zhang and H. Zhu, HoMonit: Monitoring smart home apps from encrypted traffic, *Proceedings of the ACM SIGSAC Conference on Computer and Communications Security,* pp. 1074–1088, 2018.

[46] Y. Zhang and J. Chen, Modeling virtual channel to enforce runtime properties for IoT services, *Proceedings of the Second International Conference on the Internet of Things, Data and Cloud Computing,* article no. 102, 2017.

V

IMAGE FORENSICS

Chapter 12

INDIAN CURRENCY DATABASE FOR FORENSIC RESEARCH

Saheb Chhabra, Gaurav Gupta, Garima Gupta and Monika Gupta

Abstract   Criminals are always motivated to counterfeit currency notes, especially higher denomination notes. Low-quality counterfeits are created using high-resolution scanners and printers whereas high-quality counterfeits are created using sophisticated currency printing presses and raw materials, often with the assistance of hostile nation states. Identifying counterfeit currency notes is a challenging problem that is hindered by the absence of a publicly-available database of genuine and counterfeit currency notes due to legal constraints. On November 8, 2016, the Government of India declared all 500 and 1,000 denomination notes of the Mahatma Gandhi Series as invalid tender. This research was able to collect and investigate genuine and counterfeit versions of the demonetized notes. Several new security features in the demonetized currency notes were identified and a database of microscope and scanner images has been created for forensic research.

Keywords: Counterfeit currency, security features, database

1.   Introduction

Rapid advancements in printing and scanning technologies have simplified the task of creating fake documents. Creating counterfeit currency, especially high denomination notes, is on the rise. Low-quality counterfeit currency is easily produced using high-resolution scanners and printers. In contrast, producing high-quality counterfeit currency requires sophisticated printing presses and specialized raw materials. Access to printing equipment and raw materials are next to impossible for individuals, which is why high-quality counterfeits are typically created with the assistance of hostile nation states.

Over more than 150 years, sovereign nations have taken strong steps to prevent the creation and use of counterfeit currency. On November

© IFIP International Federation for Information Processing 2021
Published by Springer Nature Switzerland AG 2021
G. Peterson and S. Shenoi (Eds.): Advances in Digital Forensics XVII, IFIP AICT 612, pp. 237–253, 2021.
https://doi.org/10.1007/978-3-030-88381-2_12

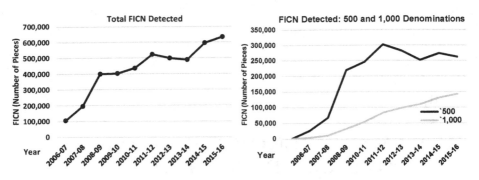

Figure 1. Detection of fake Indian currency notes by year.

8, 2016, the Government of India declared 500 and 1,000 denomination notes of the Mahatma Gandhi Series as invalid tender. This was done to combat the massive amounts of "black money" (i.e., unreported income) and to flush out counterfeit currency. The same day, India released new 500 and 2,000 denomination currency notes, claiming that their security features made it next to impossible to make counterfeits.

However, shortly after the release, counterfeit currency notes with the new 500 and 2,000 denominations were seized at the India-Bangladesh border [13]. Forensic analysis revealed that the seized counterfeit currency was printed on official Bangladesh stamp paper. Moreover, 11 of the 17 security features of the new currency notes were replicated [13], indicating that criminal entities can quickly produce high-quality counterfeit currency notes.

According to the Reserve Bank of India (India's Central Bank) [16], the number of fake Indian currency notes (FICN) detected increased from 104,743 in 2006-2007 to 632,926 in 2015-2016, an increase of more than 500% over ten years. Additionally, the number of counterfeit 1,000 denomination notes that were seized increased by more than 164% over the six-year period from 2010-2011 to 2015-2016. Figure 1 shows the total numbers of fake currency notes detected annually (left-hand side) and the numbers of fake 500 and 1,000 denomination notes detected annually (right-hand side).

Information about anti-counterfeiting technologies, including security features such as intaglio printing, micro-printing, magnetic ink character recognition (MICR), optically-variable ink (OVI) and watermarks, is readily available on the Internet. Since this information is leveraged by counterfeiters to design counterfeit currency, it is important to identify new security features in currency notes that can forensically verify their authenticity.

With input from forensic experts, this research has identified new security features that have the potential to distinguish between genuine and counterfeit currency notes with high accuracy. To advance these efforts, a database of demonetized Indian 500 and 1,000 denomination currency notes has been created. The database comprises 6,173 images of genuine and counterfeit currency notes captured using a microscope and/or scanner. The currency samples cover every year from pre-2005 to 2013 and incorporate the different inset letters. Indeed, this database of currency notes is the first of its kind to be created for forensic research and to advance the development of automated systems for currency authentication and counterfeit currency detection.

2. Related Work

Researchers have proposed several techniques for distinguishing counterfeit currency from genuine currency. Frosini et al. [8] proposed a neural-network-based recognition and verification technique that uses an inexpensive optoelectronic sensor to check banknote dimensions; this information is combined with a multilayer perceptron that recognizes the orientation and obverse side of banknotes. Other techniques for recognizing currency notes are based on size, color and texture [1, 10].

Principal component analysis and wavelet transformation techniques have been used to recognize single-country as well as multi-country banknotes [2, 6, 14]. Gai et al. [9] have proposed a scanner-based system for banknote classification based on the quaternion wavelet transform. The transform is applied to banknote images for feature extraction, which yields a shift-invariant magnitude and three phases for each sub-band. Generalized Gaussian density is then used to obtain the parameter means and standard deviations. The feature vectors are input to a backpropagation neural network for classification. This system has been tested on U.S. dollars, Chinese renminbi and European Union euros.

Other techniques for currency recognition leverage edge detection, region properties and similarity measurements [3, 11, 19]. Lee et al. [12] have designed a paper-based chipless tag system for recognizing banknotes. Xie et al. [20] compute texture roughness from Chinese renminbi currency note images to identify counterfeits. They claim that the surface of the printing layer in genuine notes tends to be rougher than in counterfeit notes; eight parameters are computed, four parameters represent statistical information whereas the other four express characteristics of gray-scale profiling.

Other researchers have extracted size, color and texture features to authenticate currency notes [4, 15, 21]. Roy et al. [18] have developed a system to authenticate paper currency, including Indian banknotes. Security features such as printing techniques, ink properties, security threads and artwork were evaluated. The system has been tested on genuine and counterfeit samples of Indian 500 denomination banknotes. In other work, Roy et al. [17] have proposed a technique for extracting gray-level, color and texture features to recognize and authenticate Indian currency notes.

Bozicevic et al. [5] have developed a non-destructive, micro-Raman spectroscopy method to detect color-printed counterfeit notes. They analyzed cyan, magenta and yellow toner cartridges from one manufacturer against a printer and toner cartridges from another manufacturer. The analysis indicated that the cyan toners have similar spectra, but there are considerable differences for the other colors.

Cozzella et al. [7] encode information about the positions of random security fibers that glow under ultra-violet light and create 2D barcodes with the encoded information. The authenticity of a banknote is verified by passing an image taken under ultra-violet light and comparing barcodes.

Research has primarily focused on authenticating currency notes in non-forensic contexts. The principal reason is the lack of a currency database that supports forensic research. The research described in this chapter has sought to collect and investigate genuine and counterfeit versions of currency notes. Several new security features in the demonetized currency notes have been identified and a database of microscope and scanner images has been created for forensic research.

3.   Indian Currency Security Features

Numerous security features are incorporated in currency notes to combat forgery and verify authenticity. Countries have unique currency designs, security features, colors, sizes and paper types. Indian currency is printed by the Reserve Bank of India. Details about the paper and raw materials used to print Indian currency notes are kept secret. Vendors are forbidden from selling them to any other entity.

Printing currency notes involves several stages. Some security features such as watermarks and security threads are incorporated when the paper is manufactured. Other features such as guilloche patterns, micro-letters and intaglios are incorporated using sophisticated printing technologies. Features such as security fibers and fluorescent ink are hidden and are only visible when illuminated by ultra-violet light.

In general, currency security features are categorized into three levels based on their complexity and protections against counterfeiting.

Figure 2. Indian currency note with intaglio security features.

First-level features such as watermarks, color-shifting inks and security inks are designed to enable individuals to verify the authenticity of currency notes. Second-level security features such as guilloche patterns and micro-printing are intended to counter forgery and are visible under magnification. Third-level features such as fluorescent inks, printing inks and security fibers are difficult to replicate and typically provide the ground-truth for authentication. All the security features must be considered when conducting forensic examinations of currency notes.

Figure 2 shows images of an Indian 500 denomination currency note and its security features. Figure 2(a) shows the obverse side of the note and Figure 2(b) shows detailed images of the intaglio security features.

Before the November 8, 2016 demonetization, valid Indian currency notes included the 5, 10, 20, 100, 500 and 1,000 denominations. After the old 500 and 1,000 denominations were declared invalid, new designs and series of 500 and 2,000 denomination currency notes were released. The Reserve Bank of India currently prints Indian currency notes at presses in four cities. Each printing press is assigned specific inset letters to be printed on currency notes. Table 1 shows the mappings of inset letters to currency printing presses.

Table 1. Mappings between inset letters and currency printing presses.

Inset Letters	Printing Press
Nil, A, B, C, D	Press 1
E, F, G, H, K	Press 2
L, M, N, P, Q	Press 3
R, S, T, U, V	Press 4

4. Indian Currency Database

Creating the Indian currency database comprising genuine and counterfeit 500 and 1,000 denomination notes involved three phases: (i) sample collection, (ii) security feature identification and (iii) database creation.

4.1 Sample Collection

Collecting counterfeit currency samples is always a challenging task. The primary barriers are legal constraints that restrict access to and collection of seized counterfeit currency. The invalidation of Indian 500 and 1,000 denomination notes enabled the collection of genuine and high-quality counterfeit samples. A total of 599 currency samples were collected, 464 genuine and 135 counterfeit notes.

The year that an Indian currency note is printed is recorded on the reverse side of the note and a designated inset letter corresponding to the printing press is printed in the background of the number panel. It was relatively easy to collect genuine currency notes printed every year from pre-2005 through 2013 that covered all the designated inset letters. To make the database more realistic, genuine currency note samples of varying quality (e.g., lightly used, heavily used and very heavily used) were included. The counterfeit currency samples were obtained from law enforcement and other agencies only after providing assurances that the database would be used exclusively for forensic research.

4.2 Security Feature Identification

The Reserve Bank of India issues guidelines for distinguishing genuine Indian currency notes from counterfeit notes. However, some of the guidelines are unreliable for identifying counterfeits. Manual analysis of genuine and counterfeit currency notes was conducted using a microscope. The analysis revealed several new security features that could be used to distinguish between genuine and counterfeit currency. In

Table 2.   Security features identified in 500 and 1,000 denomination notes.

500 Denomination Notes		
Watermark lines	See-through	Braille marks
Emblem	Number panel 2	Governor in Hindi
Flower	Governor in English	Mouth
Eye	Tiger	Latent
Micro-letters	500 on obverse	Number panel 1
Security thread 1	Security thread 2	Optically-variable ink
R	Paper fibers	Language panel
Gandhi	500 on reverse	Floral design
Lady		
1,000 Denomination Notes		
Watermark lines	See-through	Braille marks
Emblem	Number panel 2	Governor in Hindi
Flower	Governor in English	Mouth
Eye	Tiger	Latent
Micro-letters	1000 on obverse	Number panel 1
Security thread 1	Security thread 2	Optically-variable ink
R	Paper fibers	Language panel
Satellite	1000 on reverse	Floral design
Girl		

fact, 25 potential security features were identified in the 500 and 1,000 denomination notes in the currency database.

Table 2 lists the security features identified in the 500 and 1,000 denomination notes. Some of the security features are mentioned in the Reserve Bank of India guidelines whereas others were identified via the manual analysis of currency notes.

Figures 3(a) and 3(b) show the labeled security features on the obverse and reverse sides of 500 and 1,000 denomination notes, respectively.

4.3   Database Creation

The database comprises images of genuine and counterfeit 500 and 1,000 denomination notes captured using a microscope or scanner. An ISM PM200SB digital microscope was used to capture images of security features in the currency note samples. The microscope images were captured at magnifications from 10x to 200x under light intensities ranging from 2,500 to 3,000 lux; two Phillips lamps were to used obtain the required light intensities. A Canon 9000F Mark II scanner was used to capture images of the obverse and reverse of currency note samples at a resolution of 1,200 dpi. A total of 6,173 images were captured.

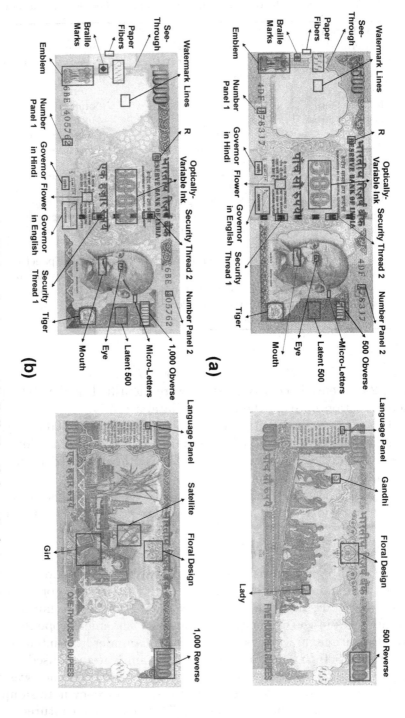

Figure 3. 500 and 1,000 denomination notes with all the identified security features.

Table 3. Dataset A details.

Currency Note Type	Samples	Microscope Images	Scanner Images
500 Genuine	118	118 × 25	118 × 2
500 Counterfeit	27	27 × 25	27 × 2
1,000 Genuine	46	46 × 25	46 × 2
1,000 Counterfeit	8	8 × 25	8 × 2
Total	199	4,975	398

The database is divided into Dataset A and Dataset B. Dataset A comprises microscope and scanner images whereas Dataset B comprises only scanner images.

Dataset A. Dataset A comprises images of the 25 security features of currency note samples captured using the microscope, and complete obverse and reverse images of currency note samples captured using the scanner. Dataset A contains a total of 5,373 images corresponding to 199 currency note samples.

Figure 4 shows the labeled security feature images of genuine (left-hand side images) and counterfeit (right-hand side images) 500 denomination notes, respectively. All the images were captured using the microscope.

Figure 5 shows the labeled security feature images of genuine (left-hand side images) and counterfeit (right-hand side images) 1,000 denomination notes, respectively. All the images were captured using the microscope.

Table 3 provides details about the images in Dataset A.

Table 4 provides details about the genuine 500 and 1,000 denomination note images in Dataset A. Dataset A contains images of genuine currency notes printed each year from pre-2005 through 2013. The currency note images collectively cover all the inset letters that were printed during each year.

Dataset A also includes images of genuine currency notes with different levels of wear and tear (i.e., lightly used, heavily used and very heavily used samples). Note that the microscope images cover all 25 security features for each sample and the scanner images cover the obverse and reverse sides of each sample. For example, in the case of the 13 genuine 500 denomination note samples from 2006, Dataset A con-

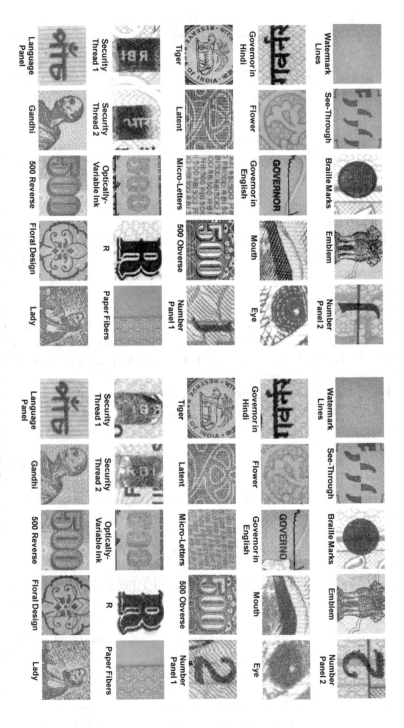

Figure 4. Security features of genuine and counterfeit 500 denomination notes.

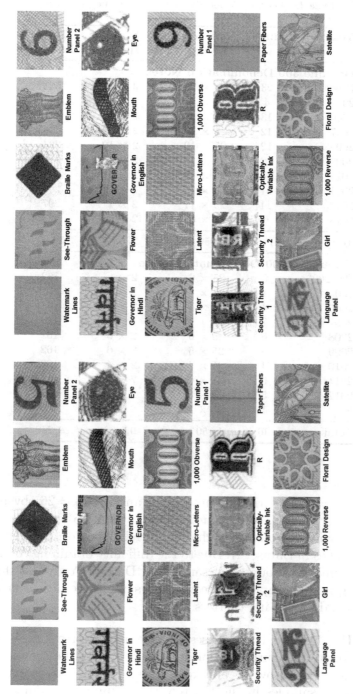

Figure 5. Security features of genuine and counterfeit 1,000 denomination notes.

Table 4.   Dataset A details (genuine currency notes).

Year	Samples	Microscope Images	Scanner Images	Total Images
Genuine 500 Denomination Notes				
Pre-2005	23	25 × 23	2 × 23	621
2005	4	25 × 4	2 × 4	108
2006	13	25 × 13	2 × 13	351
2007	12	25 × 12	2 × 12	324
2008	12	25 × 12	2 × 12	324
2009	12	25 × 12	2 × 12	324
2010	14	25 × 14	2 × 14	378
2011	11	25 × 11	2 × 11	297
2012	11	25 × 11	2 × 11	297
2013	6	25 × 6	2 × 6	162
Total	118	2,950	236	3,186
Genuine 1,000 Denomination Notes				
Pre-2005	6	25 × 6	2 × 6	162
2005	1	25 × 1	2 × 1	27
2006	4	25 × 4	2 × 4	108
2007	7	25 × 7	2 × 7	189
2008	6	25 × 6	2 × 6	162
2009	6	25 × 6	2 × 6	162
2010	4	25 × 4	2 × 4	108
2011	6	25 × 6	2 × 6	162
2012	5	25 × 5	2 × 5	135
2013	1	25 × 1	2 × 1	27
Total	46	1,150	92	1,242

tains 25 × 13 = 325 security feature images and 2 × 13 = 26 obverse and reverse scanner images, corresponding to a total of 351 images.

Table 5 provides details about the counterfeit 500 and 1,000 denomination note images in Dataset A. Due to the difficulty obtaining counterfeit notes, the number of counterfeit samples in Dataset A are less than the number of genuine samples. Moreover, the samples do not cover all the print years and all the inset letters.

Dataset B. Dataset B comprises obverse and reverse images of genuine and counterfeit currency notes captured using the scanner. Figure 6(a) shows the obverse and reverse images of 500 denomination notes (genuine samples in the top row and counterfeit samples in the bottom row). Likewise, Figure 6(b) shows the obverse and reverse im-

Figure 6. Scanned obverse and reverse images of genuine and counterfeit notes.

Table 5. Dataset A details (counterfeit currency notes).

Year	Samples	Microscope Images	Scanner Images	Total Images
Counterfeit 500 Denomination Notes				
Pre-2005	18	25 × 18	2 × 18	486
2007	2	25 × 2	2 × 2	54
2008	3	25 × 3	2 × 3	71
2010	4	25 × 4	2 × 4	108
Total	27	675	54	729
Counterfeit 1,000 Denomination Notes				
Pre-2005	2	25 × 2	2 × 2	54
2007	2	25 × 2	2 × 2	54
2010	4	25 × 4	2 × 4	108
Total	8	200	16	216

ages of 1,000 denomination notes (genuine samples in the top row and counterfeit samples in the bottom row).

Table 6. Dataset B details (genuine and counterfeit currency notes).

Currency Note Type	Samples	Scanner Images
Genuine 500 Denomination (Series 1 to 100)	100	200
Genuine 1,000 Denomination (Series 1 to 100)	100	200
Genuine 1,000 Denomination (Mixed)	100	200
Counterfeit 1,000 Denomination	100	200
Total	400	800

Table 6 provides details about the genuine and counterfeit 500 and 1,000 denomination note images in Dataset B. Specifically, Dataset B includes the obverse and reverse images of 100 samples each of genuine 500 and 1,000 denomination notes in series (i.e., the two sets of notes are in exact serial order from 1 to 100). Also, Dataset B includes the obverse and reverse images of 100 samples of genuine 1,000 denomination notes in mixed order (i.e., from different series) and 100 samples of counterfeit 1,000 denomination notes.

5.   Conclusions

Rapid advancements in printing and scanning technologies have simplified the task of creating counterfeit currency. Low-quality counterfeit currency is easily produced using a high-resolution scanner and printer. In contrast, producing high-quality counterfeit currency requires sophisticated currency printing presses and specialized raw materials that are only available to nation states. Nevertheless, high-quality counterfeits abound. Shortly after the Government of India released new 500 and 2,000 denomination currency notes with 17 advanced security features on November 8, 2016, high-quality counterfeits were seized at the India-Bangladesh border. The seized currency was printed on official Bangladesh stamp paper and 11 of the 17 security features were replicated, indicating that criminal entities can rapidly produce high-quality counterfeit currency notes.

Counterfeit currency is a serious problem that must be addressed. Security features specified in the guidelines for authenticating genuine notes are not reliable. This research has identified 25 security features in demonetized Indian 500 and 1,000 denomination currency notes that can potentially be used to distinguish between genuine and counterfeit currency notes. It has also created a database comprising 6,173 microscope and scanner images of genuine and counterfeit currency notes in the 500 and 1,000 denominations. This database is the first of its kind to be created for forensic research with the goal of spurring the development of automated systems for currency authentication.

References

[1] A. Abbasi, A review of different currency recognition systems for Bangladesh, India, China and Euro currency, *Research Journal of Applied Sciences, Engineering and Technology*, vol. 7(8), pp. 1688–1690, 2014.

[2] A. Ahmadi, S. Omatu, T. Fujinaka and T. Kosaka, Improvement of reliability in banknote classification using reject option and local PCA, *Information Sciences*, vol. 168(1-4), pp. 277–293, 2004.

[3] A. Ali and M. Manzoor, Recognition system for Pakistani paper currency, *Research Journal of Applied Sciences, Engineering and Technology*, vol. 6(16), pp. 3078–3085, 2013.

[4] R. Bhavani and A. Karthikeyan, A novel method for counterfeit banknote detection, *International Journal of Computer Sciences and Engineering*, vol. 2(4), pp. 165–167, 2014.

[5] M. Bozicevic, A. Gajovic and I. Zjakic, Identifying a common origin of toner-printed counterfeit banknotes by micro-Raman spectroscopy, *Forensic Science International*, vol. 223(1-3), pp. 314–320, 2012.

[6] E. Choi, J. Lee and J. Yoon, Feature extraction for bank note classification using wavelet transforms, *Proceedings of the Eighteenth International Conference on Pattern Recognition*, pp. 934–937, 2006.

[7] L. Cozzella, C. Simonetti and G. Spagnolo, Is it possible to use biometric techniques as authentication solutions for objects? Biometry vs. hylemetry, *Proceedings of the Fifth International Symposium on Communications, Control and Signal Processing*, 2012.

[8] A. Frosini, M. Gori and P. Priami, A neural-network-based model for paper currency recognition and verification, *IEEE Transactions on Neural Networks*, vol. 7(6), pp. 1482–1490, 1996.

[9] S. Gai, G. Yang and W. Minghua, Employing quaternion wavelet transform for banknote classification, *Neurocompuing*, vol. 118, pp. 171–178, 2013.

[10] H. Hassanpour, A. Yaseri and G. Ardeshiri, Feature extraction for paper currency recognition, *Proceedings of the Ninth International Symposium on Signal Processing and its Applications*, 2007.

[11] V. Jain and R. Vijay, Indian currency denomination identification using an image processing technique, *International Journal of Computer Science and Information Technologies*, vol. 4(1), pp. 126–128, 2013.

[12] W. Lee, H. Jang, K. Oh and J. Yu, Design of chipless tag with electromagnetic code for paper-based banknote classification, *Proceedings of the Asia-Pacific Microwave Conference*, pp. 1406–1409, 2011

[13] S. Mishra, Pakistan finds a way to counterfeit "high security" Rs. 2,000 currency notes; 11 out of 17 security features copied in fake currency, *India.Com*, February 13, 2017.

[14] J. Ok, C. Lee, E. Choi and Y. Baek, Fast country classification of banknotes, *Proceedings of the Fourth International Conference on Intelligent Systems, Modeling and Simulation*, pp. 234–236, 2013.

[15] M. Rahmadhony, S. Wasista and E. Purwantini, Validity currency detector with optical sensor using backpropagation, *Proceedings of the International Electronics Symposium*, pp. 257–262, 2015.

[16] Reserve Bank of India, Annual Report 2020-2021, Mumbai, India (`www.rbi.org.in/scripts/AnnualReportPublications.aspx?Id=1181`), 2021.

[17] A. Roy, B. Halder and U. Garain, Authentication of currency notes through printing technique verification, *Proceedings of the Seventh Indian Conference on Computer Vision, Graphics and Image Processing*, pp. 383–390, 2010.

[18] A. Roy, B. Halder, U. Garain and D. Doermann, Machine-assisted authentication of paper currency: An experiment on Indian banknotes, *International Journal on Document Analysis and Recognition*, vol. 18(3), pp. 271–285, 2015.

[19] N. Semary, S. Fadl, M. Essa and A. Gad, Currency recognition system for the visually impaired: Egyptian banknotes as a case study, *Proceedings of the Fifth International Conference on Information and Communications Technology and Accessibility*, 2015.

[20] J. Xie, C. Qin, T. Liu, Y. He and M. Xu, A new method to identify the authenticity of banknotes based on texture roughness, *Proceedings of the IEEE International Conference on Robotics and Biomimetics*, pp. 1268–1271, 2009.

[21] W. Yan and J. Chambers, An empirical approach for digital currency forensics, *Proceedings of the IEEE International Symposium on Circuits and Systems*, pp. 2988–2991, 2013.

Chapter 13

SECURITY AND PRIVACY ISSUES RELATED TO QUICK RESPONSE CODES

Pulkit Garg, Saheb Chhabra, Gaurav Gupta, Garima Gupta and Monika Gupta

Abstract   Quick response codes are widely used for tagging products, sharing information and making digital payments due to their robustness against distortion, error correction features and small size. As the adoption and popularity of quick response codes have grown, so have the associated security and privacy concerns. This chapter discusses advancements in quick response codes and the major security and privacy issues related to quick response codes.

Keywords: Quick response codes, security, privacy

1.   Introduction

During the 1940s, the growing goods and supply chain management industry sought an automated means for tracking goods and maintaining inventories. At the same time, there was a drive to do away with cash registers that required merchants to enter prices manually. The barcode, invented by Woodland and Silver [37] in 1951, provided effective solutions to these problems. Products were labeled with barcodes that were read by optical sensors. The barcodes worked seamlessly with the point-of-sale systems to log products, their prices and other attributes.

Early barcodes had black and white one-dimensional bars that encoded product information. Their popularity led to demands for more storage capacity, increased robustness and smaller size. However, increasing the storage capacity increased barcode size. To address this limitation, multi-dimensional barcodes were created that increased the processing speed by allowing them to be read in multiple directions concurrently. The first two-dimensional barcode was created in 1987. This

© IFIP International Federation for Information Processing 2021
Published by Springer Nature Switzerland AG 2021
G. Peterson and S. Shenoi (Eds.): Advances in Digital Forensics XVII, IFIP AICT 612, pp. 255–267, 2021.
https://doi.org/10.1007/978-3-030-88381-2_13

was followed by various two-dimensional barcodes that comprised not just bars and boxes, but also unique characters, shapes and patterns.

In the years that followed, there was a need for quick processing barcodes that could be deployed in assembly line environments. The idea was to have a recognizable pattern in a two-dimensional barcode that would enable a scanner to identify the code quickly. Eventually, it was decided that positional information should be incorporated in the code to increase detection speed.

Researchers conducted extensive surveys of the ratios of white to black areas in pictures, posters and other printed items. The least-used ratio of black to white was found to be 1:1:3:1:1, which enabled the determination of the widths of the black and white areas in the position detection patterns. The resulting design, which could be detected rapidly from any direction, was named a quick response (QR) code.

In 1994, QR codes were formally introduced as a substitute for one-dimensional barcodes. These two-dimensional matrix codes encode information using numbers, letters and special characters. Since QR codes store information in both directions, they can store more information than one-dimensional barcodes. QR codes also have desirable characteristics such as high error correction, strong robustness against distortions and small size. At this time, there are 40 versions of QR codes that are used based on the storage and/or error correction requirements of applications.

QR codes are scanned using a mobile or handheld device with a camera/image sensor and a QR code scanning application. The captured image is first processed to extract the QR code from the image. This is accomplished by detecting three distinctive squares in the QR code and single/multiple small squares near the fourth corner. The scanner can extract a correctly-oriented QR code image; if the image is not oriented properly, the viewing angle is adjusted up to a limit to extract the QR code image. The black and white boxes in the QR code are then decoded to read the information embedded in the QR code. The QR code also has error correction data, which are error bytes that enable information recovery when the QR code has physical damage. QR codes have recovery rates of up to 30%, which makes them one of the most robust codes ever invented.

However, QR codes have certain limitations, especially with regard to security and privacy. This chapter discusses advancements in quick response codes and the major security and privacy issues related to quick response codes.

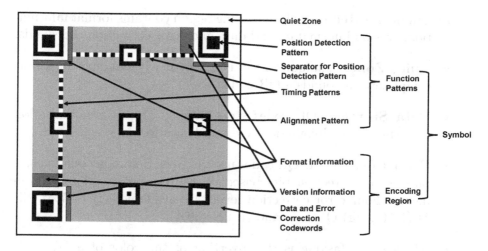

Figure 1. QR code structure [17].

2. QR Code Structure

Figure 1 presents the QR code structure. A QR code comprises black and white square modules that are set in a regular square array. The modules contain encoded information as well as the following function patterns and components:

- **Finder Pattern:** Three finder patterns are embedded in the upper-left, upper-right and lower-left corners of a QR code. These concentric square patterns assist in detecting the QR code position and orientation.

- **Separator:** A separator is an all-white pattern that is located between a finder pattern and the encoding region. The width of the separator is equal to the width of one module.

- **Timing Pattern:** A timing pattern comprises alternating black and white modules. The pattern always starts and ends with a black module. It helps determine the QR code version and identify module positions.

- **Alignment Pattern:** An alignment pattern is a 5×5 module pattern placed in a predefined position in a QR code. It comprises a 3×3 white/light module and a black/dark module in the center. The number of alignment patterns varies according to the QR code version. The first version of the QR code did not have alignment patterns.

- **Encoding Region:** The encoding region contains format information, version information and data and error correction codewords.

- **Quiet Zone:** The quiet zone is a region without markings that completely surrounds a QR code.

- **Data Storage:** QR codes store different types of information using numeric, alphabetic, byte and Kanji formats.

- **Error Correction:** QR codes use the Reed Solomon code to compute error correction bits based on the error correction requirements. Four error correction levels are available: (i) L (7%), (ii) M (15%), (iii) Q (25%) and (iv) H (30%).

- **Masking:** Masking is the inverting of the color of a QR code module. Black modules are converted to white and white modules are converted to black. Mask patterns enable QR codes to be read more easily by scanners. The color of a module after masking is determined by a formula that takes in the coordinates of the module and returns a value of zero or one. If the returned value is zero, the new color of the module is the opposite of its previous color. If the returned value is one, then the color of the module remains the same.

3.   QR Code Evolution

QR codes have evolved in three ways: (i) storage capacity, (ii) security and (iii) appearance:

- **Storage Capacity:** Researchers have proposed several solutions to increase the storage capacity of QR codes. Grillo et al. [14] proposed high-capacity colored two-dimensional codes that increase the storage capacity. Vongpradhip [35] developed a multiplexing method to increase storage capacity. Tkachenko et al. [33] suggested using a two-layered QR code in which one QR code can be read by a generic QR code scanner while the other QR code can only be read via a specialized application, thereby increasing the storage capacity and well as security and privacy. Tikhonov [32] created an innovative double-sided QR code that incorporates a QR code and its mirrored version; although this approach has some limitations, it can store two QR codes in a single QR code.

- **Security:** Since QR codes are widely used in digital transactions, government identification cards and other official documents, it is important that they are secure, non-reproducible and non-forgeable. Gaikwad and Singh [12] proposed an image embedding scheme for hiding QR codes in colored images using half-toning and luminance level control. Secret information is hidden in QR codes using steganography [4, 12]. Watermarks are used to safeguard QR codes from forgery.

- **Appearance:** A standard QR code is a matrix of black and white blocks arranged in a specific manner to store information. However, as QR codes became embedded in advertisements and hoardings, they had to be made attractive instead of just plain black and white. Chu et al. [9] proposed the use of machine-readable halftone QR codes. These QR codes are aesthetic and presentable, but have reduced error correction capabilities. Xu et al. [38] proposed a mechanism to create aesthetically pleasing QR codes while maintaining their robustness up to a certain level.

4. Key Issues

Key issues related to QR codes include their use in authentication, QR-code-based attacks, and security and privacy.

4.1 Authentication with QR Codes

Counterfeiting of documents, brands and security packaging is one of the fastest growing economic crimes. Holograms and special inks are extensively used to combat the counterfeiting of documents and products. The availability of inexpensive mobile phones, advancements in imaging technology and ease of use have led to QR codes being used to authenticate documents and products. QR codes are easy to generate and authenticate.

Lu et al. [25] proposed a secure mobile phone payment authentication scheme using visual cryptography and aesthetic QR codes. Yahya et al. [39] developed a mobile app for authenticating academic certificates. Student information is encoded in QR codes that are printed on certificates; the certificates are validated by scanning the printed QR codes. Warasart and Kuacharoen [36] implemented a solution for authenticating paper documents. A signed message is stored in a QR code that is printed on a document at creation; the integrity of the document is verified by checking the message stored in the QR code. Arkah et al. [3] created a watermarking scheme for QR codes used to authenticate electronic color documents. Keni et al. [18] developed a QR code for product authentication that is less expensive and more effective than traditional RFID-based solutions. Several other QR code designs have been developed for authenticating documents and products [2, 5, 7, 22].

4.2   Attacks Using QR Codes

QR codes can be used as attack vectors. The following attacks involving QR codes are feasible:

- **Cross-Site Scripting:** A QR code can be leveraged to execute program-based and cross-site scripting attacks. A QR code can be encoded with a URL containing an alert message that executes an exploit. When a victim accesses the URL via the QR code, the alert message executes an exploit on the victim's web browser or computer system.

- **Command Injection:** A QR code can be input as a command-line argument that enables an attacker to change the content of the QR code or attach malicious QR code over the original code. The malicious QR code can execute commands on the victim's computer system, including installing a rootkit or spyware or launching a denial-of-service attack.

- **Malware Propagation:** An attacker can encode a URL to a rogue website in a QR code. When the QR code is scanned, a connection is established to the rogue website from where malware is downloaded to the victim's computer system. In October 2011, Kaspersky Labs identified malicious sites containing QR codes for portable applications (e.g., Jimm and Opera Mini) with Trojans that sent instant messages to premium-rate numbers.

- **Malicious Pixels in QR Codes:** Kieseberg et al. [19] describe two techniques for contaminating QR code pixels for use as an attack vector. One technique involves the creation of a malicious QR code that looks similar to the original QR code. The other technique involves changing only one pixel at a time.

Attacks can also be perpetrated using barcodes that are printed with QR codes. Lee et al. [21] developed black and white two-dimensional barcodes that provide authorization using a digital signature algorithm (KCDSA). Huang et al. [16] employed a reversible data hiding technique to embed QR code in an image while preserving its integrity properties.

4.3   Security and Privacy of QR Codes

The widespread application of QR codes raises several security and privacy concerns. The easy access of information stored in QR codes is a major problem.

Considerable research has focused on using cryptography to maintain security and privacy [1, 10, 11, 15, 26, 27]. Mendhe et al. [28] proposed a three-layered QR code based message sharing system that uses a combination of cryptography and steganography. The data to be shared is encrypted using the RSA algorithm and encoded in the QR code along with a randomly-initialized pixel or mask image.

Nguyen et al. [29] designed watermark QR codes that are sensitive to printing and scanning. The technique, which does not affect readability, replaces the background of a monochrome QR code with a random texture that is clearly visible when the QR code is reproduced. Liu et al. [24] developed a secure, visual QR code that ensures the authenticity of the encoded data. A hash of a message is created and encrypted using a private key. The message is then encoded in the QR code and fused with an image. The encrypted hash is watermarked in the QR code. At the time of decoding, the hash of the encoded message is matched against the hash obtained by decrypting the watermarked string.

Zhang et al. [40] proposed a two-level QR code that stores public and private data. Public data is easily decoded using a common barcode scanner. Private data is encoded by swapping black modules in the original QR code with textured patterns. The textured patterns are sensitive to printing and scanning processes, enabling the authenticity of the QR code to be determined.

Lin et al. [23] employed steganography to prevent unauthorized access to private information stored in QR codes. The approach embeds a confidential message in the data codewords of the public message QR tag and makes no modifications to the other QR code regions. In the final QR code, the hidden information can only be extracted using a private key whereas public information is easily decoded using a normal scanner.

Krombholz et al. [20] demonstrated that malicious links to phishing sites can be embedded in QR codes. They analyzed attack scenarios in various applications of QR codes and identified design requirements for rendering QR codes secure and usable. Rogers [30] revealed a vulnerability in how Google Glass interprets QR codes. He showed that a Google Glass device could scan a QR code that would force it to connect to a hostile access point and root access could be gained to the device without the wearer's knowledge.

Interestingly, QR codes are now being placed on headstones. The QR codes typically point to websites where information, pictures and videos of the departed persons are posted. Gotved [13] discussed the privacy concerns of sharing such information in public spaces.

Bani-Hani et al. [6] discussed the privacy risks associated with QR codes. They demonstrated how sensitive user information could be stolen and how privacy rights can be violated by malicious code that runs in the background. They developed a secure system for QR code generation and scanning to address these problems. Vidas et al. [34] showed that a URL embedded in a QR code could take a smartphone user to a website that installs a malicious application that accesses and exfiltrates private user data.

5.   Innovative Applications

This section highlights three innovative applications of QR codes.

5.1   Self-Authenticating Documents

Paper document fraud is easily perpetrated using high-resolution scanners and printers. At this time, few, if any, mobile systems are available for detecting fraudulent paper documents in real time. The available systems have low accuracy, do not incorporate biometric authentication and typically require manual intervention. QR codes with security features, including biometric data, signed document content hashes and encryption, could be embedded in paper documents when they are created. A smartphone application could be used to quickly scan QR codes and verify the authenticity of the paper documents.

5.2   Color QR Codes

Limited storage capacity is limiting new applications of QR codes. Research is underway to use colors in QR codes to increase storage capacity. However, the robustness of color QR codes is an open problem.

A key hurdle is color selection. In general, the robustness of color QR codes is affected by color shifting and variations in illumination. To address these problems, QR code colors should be chosen to maximize their distances in the RGB and CMYK color spaces.

Another problem is cross-channel interference, which is the mixing of colors that occurs during the printing process. Cross-module interference is a problem that occurs when a high-density color QR code is printed using a low-resolution color printer. It affects QR code robustness because the colors tend to bleed into neighboring modules.

5.3   Anti-Counterfeiting QR Codes

QR codes are easily reproduced, replaced and forged. Current anti-counterfeiting solutions use sophisticated methods that are inefficient and not scalable. Special patterns can be employed to render QR codes non-reproducible, non-replaceable and non-forgeable, but modified QR code readers would be needed to extract, decode and verify the embedded information.

6.   Conclusions

QR codes are seeing myriad applications, but serious concerns are being raised about the security, privacy and authenticity of the information embedded in QR codes. This chapter has discussed advancements in QR code technology as well as strategies for implementing security and privacy features in QR codes. Of course, the challenge is to provide these features without reducing storage capacity and error correction capabilities.

Future research will focus on implementing anti-counterfeiting techniques for QR codes. Also, it will focus on the digital forensic aspects of copied, manipulated and forged QR codes.

References

[1] M. Ahamed and H. Mustafa, A secure QR code system for sharing personal confidential information, *Proceedings of the International Conference on Computer, Communications, Chemical, Materials and Electronic Engineering*, 2019.

[2] C. Allen and A. Harfield, Authenticating physical locations using QR codes and network latency, *Proceedings of the Fourteenth International Joint Conference on Computer Science and Software Engineering*, 2017.

[3] Z. Arkah, L. Alzubaidi, A. Ali and A. Abdulameer, Digital color document authentication using QR codes based on digital watermarking, *Proceedings of the International Conference on Intelligent Systems Design and Applications*, pp. 1093–1101, 2018.

[4] A. Arya and S. Soni, Enhanced data security in quick response (QR) code using image steganography technique with DWT-DCT-SVD, *International Journal of Scientific Research in Computer Science, Engineering and Information Technology*, vol. 3(5), pp. 59–65, 2018.

[5] E. Ayeleso, A. Adekiigbe, N. Onyeka and M. Oladele, Identity card authentication system using a QR code and smartphone, *International Journal of Science, Engineering and Environmental Technology*, vol. 2(9), pp. 61–68, 2017.

[6] R. Bani-Hani, Y. Wahsheh and M. Al-Sarhan, Secure QR code system, *Proceedings of the Tenth International Conference on Innovations in Information Technology*, 2014.

[7] A. Banu, K. Ganagavalli and G. Ramsundar, QR code based shopping with secure checkout for smartphones, *Journal of Computational and Theoretical Nanoscience*, vol. 15(5), pp. 1545–1550, 2018.

[8] T. Bui, N. Vu, T. Nguyen, I. Echizen and T. Nguyen, Robust message hiding for QR code, *Proceedings of the Tenth International Conference on Intelligent Information Hiding and Multimedia Signal Processing*, pp. 520–523, 2014.

[9] H. Chu, C. Chang, R. Lee and N. Mitra, Halftone QR codes, *ACM Transactions on Graphics*, vol. 32(6), article no. 217, 2013.

[10] Z. Cui, W. Li, C. Yu and N. Yu, A new type of two-dimensional anti-counterfeit code for document authentication using neural networks, *Proceedings of the Fourth International Conference on Cryptography, Security and Privacy*, pp. 68–73, 2020.

[11] R. Dudheria, Evaluating features and effectiveness of secure QR code scanners, *Proceedings of the International Conference on Cyber-Enabled Distributed Computing and Knowledge Discovery*, pp. 40–49, 2017.

[12] A. Gaikwad and K. Singh, Information hiding using image embedding in QR codes for color images: A review, *International Journal of Computer Science and Information Technologies*, vol. 26(1), pp. 278–283, 2015.

[13] S. Gotved, Privacy with public access: Digital memorials in quick response codes, *Information, Communication and Society*, vol. 18(3), pp. 269–280, 2015.

[14] A. Grillo, A. Lentini, M. Querini and G. Italiano, High capacity colored two-dimensional codes, *Proceedings of the International Multiconference on Computer Science and Information Technology*, pp. 709–716, 2010.

[15] V. Hajduk, M. Broda, O. Kovac and D. Levicky, Image steganography using QR code and cryptography, *Proceedings of the Twenty-Sixth International Conference on Radioelectronics*, pp. 350–353, 2016.

[16] H. Huang, F. Chang and W. Fang, Reversible data hiding with histogram-based difference expansion for QR code applications, *IEEE Transactions on Consumer Electronics*, vol. 57(2), pp. 779–787, 2011.

[17] International Organization for Standardization, ISO/IEC Standard 18004: Information Technology – Automatic Identification and Data Capture Techniques – QR Code Bar Code Symbology Specification, Geneva, Switzerland, 2000.

[18] H. Keni, M. Earle and M. Min, Product authentication using hash chains and printed QR codes, *Proceedings of the Fourth IEEE Annual Consumer Communications and Networking Conference*, pp. 319–324, 2017.

[19] P. Kieseberg, S. Schrittwieser, M. Leithner, M. Mulazzani, E. Weippl, L. Munroe and M. Sinha, Malicious pixels using QR codes as an attack vector, in *Trustworthy Ubiquitous Computing*, I. Khalil and T. Mantoro (Eds.), Atlantis Press, Paris, France, pp. 21–38, 2012.

[20] K. Krombholz, P. Fruhwirt, P. Kieseberg, I. Kapsalis, M. Huberand and E. Weippl, QR code security: A survey of attacks and challenges for usable security, *Proceedings of the International Conference on Human Aspects of Information Security, Privacy and Trust*, pp. 79–90, 2014.

[21] J. Lee, T. Kwon, S. Song and J. Song, A model for embedding and authorizing digital signatures in printed documents, *Proceedings of the International Conference on Information Security and Cryptology*, pp. 465–477, 2002.

[22] O. Lewis and S. Thorpe, Authenticating motor insurance documents using QR codes, *Proceedings of the Southeast Conference*, 2019.

[23] P. Lin and Y. Chen, QR code steganography with secret payload enhancement, *Proceedings of the IEEE International Conference on Multimedia and Expo Workshops*, 2016.

[24] S. Liu, J. Zhang, J. Pan and C. Weng, SVQR: A novel secure visual quick response code and its anti-counterfeiting solution, *Journal of Information Hiding and Multimedia Signal Processing*, vol. 8(5), pp. 1132–1140, 2017.

[25] J. Lu, Z. Yang, L. Li, W. Yuan, L. Li and C. Chang, Multiple schemes for mobile payment authentication using QR codes and visual cryptography, *Mobile Information Systems*, vol. 2017, article no. 4356038, 2017.

[26] S. Maheswari and D. Hemanth, Frequency domain QR code based image steganography using the Fresnelet transform, *AEU International Journal of Electronics and Communications*, vol. 69(2), pp. 539–544, 2015.

[27] V. Mavroeidis and M. Nicho, Quick response code secure: A cryptographically-secure anti-phishing tool for QR code attacks, *Proceedings of the International Conference on Mathematical Methods, Models and Architectures for Computer Network Security*, pp. 313–324, 2017.

[28] A. Mendhe, D. Gupta and K. Sharma, Secure QR code based message sharing system using cryptography and steganography, *Proceedings of the First International Conference on Secure Cyber Computing and Communications*, pp. 188–191, 2018.

[29] H. Nguyen, A. Delahaies, F. Retraint, D. Nguyen, M. Pic and F. Morain-Nicolier, A watermarking technique to secure printed QR codes using a statistical test, *Proceedings of the IEEE Global Conference on Signal and Information Processing*, pp. 288–292, 2017.

[30] M. Rogers, Hacking the Internet of Things for Good, Lookout, San Francisco, California (`www.slideshare.net/LookoutInc/hacking-the-internet-of-things-for-good`), 2013.

[31] J. Swartz, The growing "magic" of automatic identification, *IEEE Robotics and Automation*, vol. 6(1), pp. 20–23, 1999.

[32] A. Tikhonov, On Double-Sided QR-Codes, arXiv: 1902.05722 (`arxiv.org/abs/1902.05722`), 2019.

[33] I. Tkachenko, W. Puech, O. Strauss, C. Destruel, J. Gaudin and C. Guichard, Rich QR code for multimedia management applications, *Proceedings of the International Conference on Image Analysis and Processing*, pp. 383–393, 2015.

[34] T. Vidas, E. Owusu, S. Wang, C. Zeng, L. Cranor and N. Christin, QRishing: The susceptibility of smartphone users to QR code phishing attacks, in *Financial Cryptography and Data Security*, A. Adams, M. Brenner and M. Smith (Eds.), Springer, Berlin Heidelberg, Germany, pp. 52–69, 2013.

[35] S. Vongpradhip, Using multiplexing to increase information in QR codes, *Proceedings of the Eighth International Conference on Computer Science and Education*, pp. 361–364, 2013.

[36] M. Warasart and P. Kuacharoen, Paper-based document authentication using digital signatures and QR codes, presented at the *International Conference on Computer Engineering and Technology*, 2012.

[37] N. Woodland and B. Silver, Classifying Apparatus and Method, U.S. Patent No. 2,612,994, October 7, 1952.

[38] M. Xu, Q. Li, J. Niu, H. Su, X. Liu, W. Xu, P. Lv, B. Zhou and Y. Yang, ART-UP: A novel method for generating scanning-robust aesthetic QR codes, *ACM Transactions on Multimedia Computing, Communications and Applications*, vol. 17(1), article no. 25, 2021.

[39] Z. Yahya, N. Kamarzaman, N. Azizan, Z. Jusoh, R. Isa, M. Shafazand, N. Salleh and S. Mokhtaruddin, A new academic certificate authentication using leading-edge technology, *Proceedings of the International Conference on E-Commerce, E-Business and E-Government*, pp. 82–85, 2017.

[40] B. Zhang, K. Ren, G. Xing, X. Fu and C. Wang, SBVLC: Secure barcode-based visible light communications for smartphones, *IEEE Transactions on Mobile Computing*, vol. 15(2), pp. 432–446, 2016.

Printed in the United States
by Baker & Taylor Publisher Services